WH250 BAI

24 0421042 8

A

AF845274

The Library

Imperial College School of Medicine at St. Mary's

This book should be returned to the library on or before the last date below. Books may be renewed on application to the library either in person or by telephoning: 0171 594 3692

~~01 NOV 2000~~
-9 NOV 2001
~~10 FEB 2003~~
~~20 FEB 2003~~
-4 JAN 2004
20 DEC 2005
22 DEC 2005
18 DEC 2007
21 DEC 2009
25 JAN 2011
16 MAY 2011
20 DEC 2011

WITHDRAWN

ital Campus

dicine

gy Laboratory

Medical Intelligence Unit

Chromosomes and Genes in Acute Lymphoblastic Leukemia

Lorna M. Secker-Walker, MA, PhD, FRCPath

Royal Free Hospital, School of Medicine
London, England

THE LIBRARY
IC SCHOOL OF MEDICINE
AT ST. MARY'S

Springer
New York Berlin Heidelberg London Paris
Tokyo Hong Kong Barcelona Budapest

R.G. Landes Company
Austin

Medical Intelligence Unit

CHROMOSOMES AND GENES IN ACUTE LYMPHOBLASTIC LEUKEMIA

R.G. LANDES COMPANY
Austin, Texas, U.S.A.

International Copyright © 1997 Springer-Verlag, Heidelberg, Germany

All rights reserved.
No part of this book may be reproduced or transmitted in any form or by any means, electronic or mechanical, including photocopy, recording, or any information storage and retrieval system, without permission in writing from the publisher.
Printed in the U.S.A.

Please address all inquiries to the Publishers:
R.G. Landes Company, 810 S. Church Street, Georgetown, Texas, U.S.A. 78626
Phone: 512/ 863 7762; FAX: 512/ 863 0081

International distributor (except North America):
Springer-Verlag GmbH & Co. KG
Tiergartenstrasse 17, D-69121 Heidelberg, Germany

17660/HGJ

Springer

THE LIBRARY
IC SCHOOL OF MEDICINE
AT ST. MARY'S

International ISBN: 3-540-61969-0

While the authors, editors and publisher believe that drug selection and dosage and the specifications and usage of equipment and devices, as set forth in this book, are in accord with current recommendations and practice at the time of publication, they make no warranty, expressed or implied, with respect to material described in this book. In view of the ongoing research, equipment development, changes in governmental regulations and the rapid accumulation of information relating to the biomedical sciences, the reader is urged to carefully review and evaluate the information provided herein.

Library of Congress Cataloging-in-Publication Data

Secker-Walker, Lorna M.
Chromosomes and genes in acute lymphoblastic leukemia / Lorna M. Secker-Walker
p. cm. — (Medical intelligence unit)
Includes bibliographical references and index.
ISBN 1-57059-411-2 (alk. paper)
1. Lymphoblastic leukemia—Genetic aspects. I. Title. II. Series.
[DNLM: 1. Leukemia, Lymphocytic, Acute—genetics. 2. Chromosome Abnormalities.
WH 250 S444c 1996]
RC643.S43 1996
616.99'419042—dc20
DNLM/DLC 96-43702
for Library of Congress CIP

Publisher's Note

R.G. Landes Company publishes six book series: *Medical Intelligence Unit, Molecular Biology Intelligence Unit, Neuroscience Intelligence Unit, Tissue Engineering Intelligence Unit, Biotechnology Intelligence Unit* and *Environmental Intelligence Unit.* The authors of our books are acknowledged leaders in their fields and the topics are unique. Almost without exception, no other similar books exist on these topics.

Our goal is to publish books in important and rapidly changing areas of bioscience and environment for sophisticated researchers and clinicians. To achieve this goal, we have accelerated our publishing program to conform to the fast pace in which information grows in bioscience. Most of our books are published within 90 to 120 days of receipt of the manuscript. We would like to thank our readers for their continuing interest and welcome any comments or suggestions they may have for future books.

Shyamali Ghosh
Publications Director
R.G. Landes Company

Dedication

To my husband David.

CONTENTS

FOREWORD

For many years after the discovery of the Philadelphia chromosome in 1960, this and other clonal chromosomal abnormalities detected in patients with hematological diseases were considered to be curiosities with little, if any, relevance to the etiology of these diseases or to their management. In 1978, Lorna Secker-Walker, Sylvia Lawler and Roger Hardisty first showed that the chromosomal findings at diagnosis gave prognostic information in children with acute lymphoblastic leukemia. This pivotal observation introduced two decades in which chromosome analysis has become increasingly important in the diagnosis and management of the acute leukemias. Detailed study of these chromosomal changes at a molecular level has revealed the genes involved in the malignant transformation. This has increased the importance of chromosome analysis and extended its contribution to patient management.

The present book, written by one of the world's leading cytogeneticists in the field of hematology, describes the large numbers of chromosome changes that have now been reported in acute lymphoblastic leukemia (ALL). Both the well established method of banding and the newer fluorescence in situ hybridization, Southern blotting and the polymerase chain reaction techniques are reviewed. A glossary of terms used in this field, essential for all nonchromosomal cogniscenti, is included. The most important translocations, partial deletions and chromosome gains and losses are illustrated and the clinical and molecular significance of each is described.

Lorna Secker-Walker's earlier research work was performed at the Royal Marsden Hospital. However, since 1984 she has developed, in the Department of Haematology at the Royal Free Hospital and School of Medicine, a major Cytogenetic Unit in the field of hematological malignancies. Here she has made important contributions in the discovery of new translocations and establishment of their significance at clinical and molecular levels. Lorna is widely recognized internationally both for her research achievements and for the training of many workers in the cytogenetic field, and has raised standards of chromosomal analysis internationally.

It has been a privilege to be associated with Lorna and her work in the cytogenetic and molecular abnormalities in the hematological malignancies. It is now a great honor to write a Foreword to this gem of a book which all clinicians and basic 'scientists working in the field' will wish to own and read.

A. Victor Hoffbrand, MA, DM, FRCP, FRCPath, FRCP (Edin), DSc
Department of Haematology
Royal Free Hospital and School of Medicine
July 5, 1996

Acknowledgments

I would like to thank all colleagues with whom I have worked on the cytogenetics of malignancy. In particular, I should like to mention Professor Humphrey Kay, the late Professor Sylvia D. Lawler and Professor A.V. Hoffbrand for help and encouragement over many years. I also wish to acknowledge the contribution of past and present members of my group who have assisted in many of the research projects referred to in this book. Special thanks are due to those of my present team who have prepared karyograms and diagrams for publication. I am grateful to my members of staff on the LRF/UKCCG karyotype database, to the members of the UKCCG and to the Coordinators and Working parties of the MRC trials UKALLXI and XII for permission to include some of the findings from the database in Appendix 2.

SYMBOLS AND ABBREVIATIONS

add	Additional material of unknown origin.
brackets, angle <>	Used to indicate ploidy level.
brackets,square []	Used to indicate the number of cells with a clonal abnormality.
c	Constitutional anomaly, e.g., 47,XY, +21c Trisomy 21 is constitutional.
comma(,)	Separates chromosome numbers, sex chromosomes, and chromosome abnormalities.
cp	Composite karyotype. Each abnormality has been seen in at least two cells but there may be no cell with all the abnormalities.
dash (-)	Uncertain breakpoint localization.
del	Deletion. Denotes both terminal and interstitial deletion.
der	Derivative chromosome—one of the structurally rearranged chromosomes generated by a rearrangement involving two or more chromosomes.
dic	Dicentric chromosome having two centromeres, made of up two chromosomes.
dmin	Double minute chromosome—an acentric marker.
dup	Duplication of part of a chromosome.
hsr	Homogeneously staining region of a chromosome.

i	Isochromosome—loss of short or long arm, and duplication of remaining long or short arm.
idem	Used to denote the stemline karyotype in subclones.
ider	Isoderivative chromosome—isochromosome of a derivative chromosome.
idic	Isodicentric chromosome
inc	Incomplete karyotype because of poor chromosome quality, unidentified changes additional to those listed were seen.
ins	Insertion of part of one chromosome into another.
inv	Inversion of part of a chromosome.
mar	Marker chromosome - one in which no part can be identified.
minus (-)	Loss.
multiplication sign (x)	Denotes multiple copies of clones or rearranged chromosomes.
or	Alternative interpretation.
p	Short arm of chromosome.
p10	Short arm part of the centromere.
parenthesis ( )	Surrounds structurally altered chromosome(s) and breakpoints.

Ph	Philadelphia chromosome.
plus (+)	Gain.
q	Long arm of chromosome.
q10	Long arm part of the centromere.
question mark (?)	Questionable identification of the following chromosome or chromosome structure.
r	Ring chromosome.
semicolon (;)	Separates chromosomes and bands in structural rearrangements involving more than one chromosome.
slant line(/)	Separates clones.
t	Translocation.
ter	Terminal bands of a chromosome.
underline	Distinguishes homologous chromosomes.

List of symbols and abbreviated terms used to describe acquired clonal chromosome abnormalities. ISCN (1991). Guidelines for Cancer Cytogenetics, Supplement to An International System for Human Cytogenetic Nomenclature Basel: S. Karger, 1991.

CHAPTER 1

Clinical and Biological Features of Acute Lymphoblastic Leukemia

INCIDENCE, ORIGIN AND PRESENTING FEATURES

Incidence and Age

Acute lymphoblastic leukemia (ALL) is primarily a disease of childhood with 75% of patients presenting under the age of 15 years. In the United States, leukemia and lymphoma have been shown to account for 44% of all cancers among white children (<15 years of age) and for 33% of cancers among black children. The annual incidence in children 15 years or younger in the U.S. was reported to be approximately 3 cases per 100,000 for whites but no more than 1.4 cases per 100,000 for blacks. Among white children ALL is most frequent between the ages of one and five years with a peak at three to four years.[1,2] The other 25% of patients present between the ages of 15 and 80 years. Analysis by Stevens (1986) of data obtained from the Connecticut Tumor Registry for the years 1935-1982 showed that the incidence of ALL fell to a minimum by the age of 30 years and then rose sharply after the age of 40 years.[3] The epidemiology of ALL in the U.K. is similar. Between 1984 and 1986 ALL occurred in the U.K. at ages 0-84 years at an overall rate of 1.3 per 100,000 with a peak incidence between zero and four years of age (5.1 per 100,000), diminishing thereafter and becoming rare between ages 25 and 44. A slight rise was seen in people aged 75 and over (1.8 per 100,000).[4] More males than females present with ALL. The sex ratio of males:females has been given as 1.4:1.0 in childhood and said not to alter significantly in any age group.[4] The incidence was shown to be the same in males and females after adjustment for cohort effects in the Connecticut series.[3]

Origin-Predisposing Factors

The striking association between the occurrence of ALL as well as acute myeloid leukemia and certain risk groups has provided indications that the disease may have a genetic component. Siblings of leukemia patients are four times more likely to develop the disease than are the general population.[5]

Patients with a constitutional cytogenetic abnormality have a predisposition to develop lymphoid or myeloid acute leukemia. This is particularly marked for Down's syndrome patients in which the incidence of leukemia is 10- to 20-fold higher than in the general population. The peak ages of onset are in the newborn period and again at three to six years. Risk extends into adulthood.[6] There is evidence that some males with Klinefelter's syndrome with constitutional karyotype 47,XXY and individuals with a constitutional balanced translocation also have an increased risk of leukemia.[7,8] Another aspect of the constitutional karyotype which may predispose to malignancy is variations in the size of the heterochromatic regions of chromosomes 1, 9 and 16. Comparison of the heterochromatin between 67 children with ALL and 50 controls showed that the heterochromatic regions in ALL patients were significantly larger for chromosomes 1 and 9 and significantly smaller for chromosome 16 ($p < 0.001$ in each case). In addition, patients with ALL were more likely than the general population to be heteromorphic for the heterochromatic regions of chromosome 1 and 9 ($p < 0.001$).[9]

Presentation

The symptoms of ALL, including anemia, bruising, bleeding and infections, result from the lack of normal hemopoiesis. Moderate enlargement of the lymph nodes, spleen and liver are common.

A mediastinal thymic mass may be found in T ALL. Central nervous system involvement is a feature particularly in infants.[10]

LABORATORY FEATURES AND CLASSIFICATION

ALL, far from being a single disease entity, is a diverse group of hematological diseases composed of many recognizable subtypes.

WHITE BLOOD CELL COUNT

The number of white blood cells (WBC) at diagnosis ranges from <5 to >100 x 10^9/L. With WBCs greater than 5 x 10^9/l leukemic blasts are found in the peripheral blood. For practical purposes classification divides patients by WBC as: <10, 10-20 or 10-50, 50-100 and >100 x 10^9/L. Approximately 30% of children and adults have a WBC <5 x 10^9/L. Counts between 5 and 20 x 10^9/L are found in 39% of children and 36% of adults and higher counts (>20 x 10^9/L) account for 31% of children and 34% of adults. Only about 50% of patients have a WBC >10 x 10^9/L at presentation.[1,11]

BLAST CELL MORPHOLOGY: THE FRENCH-AMERICAN-BRITISH (FAB) CLASSIFICATION

FAB type is a method of defining morphological subsets of ALL. L1 describes bone marrow with small, uniform blast cells with scanty cytoplasm and moderate basophilia. Nuclei are regular in shape with inconspicuous nucleoli. L2 blasts are larger, nonuniform, with variable amounts of cytoplasm and differing degrees of basophilia. Nuclei are irregular in shape with prominent nucleoli. L3 blasts are large and uniform with moderately abundant cytoplasm, deep basophilia and prominent vacuoles. Nuclei are regular in shape with prominent nucleoli. L3 blasts are strongly associated with mature B cell immunophenotype expressing surface immunoglobulin (sIg). In some cases a gradation of blast type from L1 to L2 is seen.[12] In a childhood survey of 765 bone marrow slides undertaken in 1981, the distribution of cases between L1:L2:L3 was seen to be 85.1%:14.1%:0.8%.[13] In adults most patients are L2 type with L1:L2:L3 = 31:60:9. Classification by FAB type includes a degree of subjectivity. Differences in classification may vary between observers.

CELL LINEAGE AND IMMUNOPHENOTYPE OF THE BLASTS

The most important classification of ALL is by lineage commitment and maturation (immunophenotype) of the blasts.

Cell Lineage

Blast cell lineage can be demonstrated by investigation of rearrangement of the immunoglobulin heavy chain genes (*IGH*) (in B-lineage leukemias) or of the T cell antigen receptors (*TCR*) genes (in T ALL). However, while virtually all primitive non-T ALL cells have *IGH* rearrangements, these may also occur in primitive non-B cells. Equally, *TCR* gene rearrangements occur not only in T ALL cases but also in some non-T ALL. Thus gene rearrangements alone cannot be used to determine the cell lineage of leukemic cells.[14]

Immunophenotype

The immunophenotype of the blasts is identified by using monoclonal antibodies to cell surface antigens and cell markers. Blast cell types fall broadly into two categories: B cells and T cells. Cell type specific antigens for B cells include surface membrane immunoglobulins (sIg) and cytoplasmic immunoglobulins (cIg). Cells of the B lineage are classified by the stage of maturation from the most primitive, or most aberrant, pre-pre B or null cells (CD19+ve, CD10-ve), to early pre-B or C-ALL (CD 10+ve), to pre-B (cIg+ve CD10+ve/-ve), to mature B cells (sIg+ve). Blasts of T cell lineage have sheep erythrocyte receptors[15] and are classified as early thymocytes, CD7+ve; intermediate thymocyte CD7+, CD2+, CD1+, CD4 and/or CD8+, CyCD3; or late thymocyte CD7+, CD2+, CD1+, CD4+, surface TCRA/B T_3 complex; or mature peripheral T cell CD3+, CD4+ or CD8+ CD7+. B lineage ALL accounts for 75% to 80% of cases. C-ALL is most common in childhood (60-70% of cases), while null ALL is rare in childhood but occurs more frequently in adults (20-40%). Mature B ALL accounts for up to 5% of ALL at all age groups. Cells expressing T cell markers (T ALL) comprise 13%-20% of childhood ALL cases and up to 25% in adults.[10,15,16] Mixed lineage or bilineage leukemias, in which both lymphoid and myeloid associated surface antigens are expressed on the same cells or on different cells, respectively, in the same patient, occur in 5%-10% of patients.[17]

ETIOLOGY

ALL is a clonal disorder which arises in a single bone marrow cell which, in the majority of cases, is supposedly committed to lymphoid development. The malignant clone expands, taking over from and crowding out the normal component cells of the marrow. The expression of surface antigens of the proliferating cells, as described in the previous section, indicates cell lineage and the level of commitment of the target cell for chromosomal change. The small proportion of cases which display characteristics of more than one lineage are presumed to have arisen in a more primitive cell type uncommitted to a particular lineage.

TREATMENT

Treatment for ALL has developed steadily since the 1950s when single drug regimens were used to modest effect. Chemotherapy has become increasingly complex and the application and development of clinical trials over many years has resulted in a steady improvement in survival (see reference 10). The previous treatment trial for adults and children in the U.K. (UKALLX) consisted of induction therapy with daunorubicin, vincristine, L-asparaginase, prednisolone and intrathecal methotrexate. Drugs used for intensification were cytarabine, etoposide, thioguanine and daunorubicin. Maintenance therapy, using mercaptopurine, methotrexate, prednisolone and vincristine, was continued for two years from the time of achieving remission.[18] In the current Medical Research Council trial for adults UKALL XII, allogeneic bone marrow transplantation (BMT) is an option for all patients with a compatible donor.

PROGNOSTIC FACTORS

Biologically, ALL displays a spectrum of diseases and it is now recognized that patients can be classified as "good" or "poor" risk on the basis of a number of variables recognized at diagnosis and on their initial response to treatment. Given uniform treatment, these patients will, at one extreme, be long term survivors or, at the other, inevitably succumb to the disease within a relatively short time span. Features of the leukemia at diagnosis which are indicative of prognosis are age, white blood cell count at presentation, FAB type, immunophenotype, sex and time to remission (see Table 1.1).

Table 1.1. Factors influencing prognosis in ALL on standard therapy

	Good Prognosis	Poor Prognosis
Age yrs	2-10	<1, <2, >10
WBC x 10^9/L	<5, <10	>10, >20, >100
FAB	L1	L2 L3
Immunophenotype	C-ALL CD10+	null ALL Pre-B Mature B T ALL
Sex	Females	Males
Time to remission	<4 weeks	>4 weeks

AGE

Age is the most powerful indicator of prognosis, with infants (<1 year) having a poor prognosis, patients aged 3-5 years faring best, and patients doing increasingly less well with increasing age thereafter. This is particularly marked in adults in which survival figures become progressively worse with each decade.

WHITE BLOOD CELL COUNTS

The best prognosis is associated with white blood counts below 10 x 10^9/L. An increasingly poor prognosis is seen as WBCs rise.

FAB TYPE

Those with FAB type L1 do better than L2 cases. Children with >25% L2 blasts have a significantly higher relapse rate and poorer survival than those with >75% L1 blasts.[13] L3 cases have traditionally had a very poor outcome but with specially tailored treatment these patients may now do well.

IMMUNOPHENOTYPE

The immunophenotype of the blasts is also indicative of response to standard therapy, although this is not independent of other prognostic variables. Originally common, (C-) ALL had the best outcome, followed by pre-B, null-ALL, T ALL and B-ALL. Recognition of these prognostic factors and the development of risk-related treatment has resulted in improved survival for patients with T- or B-ALL.

SEX

Boys do less well than girls—a finding which is independent of age and presenting leukocyte count. Factors influencing prognosis were examined in over 4000 children aged 1-14 years treated on successive MRC UKALL trials from 1972 to 1990. During this time, the five year event-free survival in girls improved from 51% to 71% and in boys from 31% to 57%. These results were only partially explained by the 10% rate of testicular relapse. An excess of bone marrow relapse in boys was particularly evident after two years from diagnosis.[19]

RESPONSE TO TREATMENT

The way in which the patient responds to treatment, as determined by the clearance of blasts from the peripheral blood and bone marrow, provides another important prognostic indicator. In one study of 79 adults with ALL, multivariate analysis showed that the absence blasts cells from the peripheral blood at day 7 and the presence of <5% blasts in the bone marrow by day 28 made a significant contribution to improved event-free survival.[20]

TREATMENT

Treatment has an important impact on prognosis. For many years it was widely regarded as the single most important prognostic variable in ALL. Certainly before the 1960s, when treatment was with a single agent, the outlook for patients with "acute leukemia of the lymphoid cells" was extremely poor with median survival of 3-5 months. During the 1960s, multi-drug regimens were introduced. By 1970, 30% of children were achieving a two year disease-free survival. Today, cure is predicted for >70% of children. The outlook for adults has always been poor and even now only 30% will achieve survival of three years or longer.

CYTOGENETICS

The karyotype (chromosomal makeup) of the malignant clone is another variable diagnostic feature by which patients can be classified and has important prognostic implications. Recognition of the clinical importance of cytogenetic changes in acute lymphoblastic leukemia (ALL) dates from 1978 when their prognostic significance was first demonstrated.[21] Both chromosome number (ploidy) and chromosome structure have been identified as powerful prognostic indicators which add significance to all other variables combined.[22-25]

At a fundamental level, chromosomal abnormalities acquired by the bone marrow cells of patients with leukemia are the microscopically visible signs of submicroscopic genetic change associated with the conversion of a normal cell into a malignant one. Evidence for the central role of cytogenetic change in the etiology of this kind of leukemia is compelling. In 1983, Rowley observed that many chromosome breakpoints in the leukemias coincide with the sites of cellular oncogenes.[26] It has now become clear that chromosomal translocations bring about the unnatural juxtaposition of genes leading to the aberrant expression of a transcription factor.[27-29] Gene activation occurs in one of two ways. Most aberrant gene products result from gene fusion in which exons from normally disparate genes become joined to encode a novel fusion protein. Less common is oncogene activation brought about by the juxtaposition, but not the fusion, of a cellular oncogene with either an immunoglobulin gene (in B cell malignancies) or a T cell receptor gene (in T cell malignancies).

The impact of chromosomal change on the biology of the disease is highlighted by a number of chromosomal abnormalities which identify subgroups with remarkably similar clinical features and responses to treatment.[8,21,22,30-32] Molecular probes to a number of chromosomally induced genetic abnormalities are now available. Accordingly, the combination of cytogenetics and molecular techniques provide powerful tools with which to detect and monitor these specific markers of the disease.

In this book, I shall describe the spectrum of chromosomal findings found in ALL and the clinical features associated with them. The genetic changes brought about by chromosomal rearrangement will also be described. Alterations in the structure of genes which are not known to be associated with a chromosomal rearrangement are beyond the scope of this book. The recognition of chromosomal and chromosomally induced genetic changes have had and continue to have a profound impact on the diagnosis and management of patients with acute lymphoblastic leukemia.

REFERENCES

1. Smithson WA, Gilchrist GS, Burgert EOJ. Childhood acute lymphocytic leukemia. Cancer J Clin 1980; 30:158-181.
2. Young Jr JL, Ries LG, Silverberg E et al Cancer incidence, survival and mortality for children younger than age 15 years. Cancer 1986; 58:598-602.
3. Stevens RG. Age and risk of acute leukemia. Journal of the National Cancer Institute 1986; 76:845-848.
4. McKinney PA, Alexander FE, Cartwright RA et al. The Leukaemia Research Fund Data Collection Study: descriptive epidemiology of acute lymphoblastic leukemia. Leukemia 1989; 3:880-885.
5. Miller RW. Persons with exceptionally high risk of leukemia. Cancer Research 1967; 27:2420-2423.
6. Fong CT, Brodeur GM. Down's syndrome and leukemia: epidemiology, genetics, cytogenetics and mechanisms of leukemogenesis. [Review]. Cancer Genet Cytogenet 1987; 28:55-76.
7. Shaw MP, Eden OB, Grace E et al. Acute lymphoblastic leukemia and Klinefelter's syndrome. Pediatr Hemat Oncol 1992; 9:81-85.
8. IWCL3. Third International Workshop on Chromosomes in Leukemia (1980): Chromosomal abnormalities in acute lymphoblastic leukemia: Structural and numerical changes in 234 cases. Cancer Genet Cytogenet 1981; 4:101-110.
9. Tsezou A, Kitsiou-Tzeli S, Kosmidis H et al. Constitutive heterochromatin polymorphisms in children with acute lymphoblastic leukemia. Pediatr Hemat Oncol 1993; 10:7-11.
10. Rivera GK, Crist WM. Acute lymphoblastic leukemia. In: Handin RI, Stossel TP, Lux SE, eds. Philadelphia: J.B. Lippincott Company, 1995:743-782.
11. Mauer AM, Bowman WP, Dow LW et al. Acute lymphocytic leukemia: biological markers of clinical response. J Cell Physiol Suppl 1982; 1:127-132.
12. Bennett JM, Catovsky D, Daniel MT et al. Proposals for the classification of the acute leukaemias. French-American-British (FAB) co-operative group. Br J Haematol 1976; 33:451-458.

13. Miller DR, Leikin S, Albo V et al. Prognostic importance of morphology (FAB classification) in childhood acute lymphoblastic leukaemia (ALL). Br J Haematol 1981; 48:199-206.
14. Foon KA, Todd RF. Immunologic classification of leukemia and lymphoma. Blood 1986; 68:1-31.
15. Drexler HG, Gignac SM, Minowada J. Routine immunophenotyping of acute leukaemias. Blut 1988; 57:327-339.
16. Van der Reijden HJ, van Wering ER, van de Rijn JM et al. Immunological typing of acute lymphoblastic leukaemia. Scandinavian Journal of Haematology 1983; 30:356-366.
17. Mirro J, Zipf TF, Pui CH et al. Acute mixed lineage leukemia: clinicopathologic correlations and prognostic significance. Blood 1985; 66:1115-1123.
18. Chessells JM, Bailey C, Richards SM. Intensification of treatment and survival in all children with lymphoblastic leukaemia: results of UK Medical Research Council trial UKALL X. Medical Research Council Working Party on Childhood Leukaemia. Lancet 1995; 345:143-148.
19. Chessells JM, Richards SM, Bailey CC et al. Gender and treatment outcome in childhood lymphoblastic leukaemia: report from the MRC UKALL trials. Br J Haematology 1995; 89:364-372.
20. Legrand O, Marie JP, Cadiou M et al. Early cytoreduction: a major prognostic factor in adult acute lymphoblastic leukemia. Leukemia & Lymphoma 1994; 15:433-438.
21. Secker-Walker LM, Lawler SD, Hardisty RM. Prognostic implications of chromosomal findings in acute lymphoblastic leukaemia at diagnosis. Br Med J 1978; 2:1529-1530.
22. IWCL3 . Third International Workshop on Chromosomes in Leukemia (1980): Clinical significance of chromosomal abnormalities in acute lymphoblastic leukemia. Cancer Genet Cytogenet 1981; 4:111-137.
23. Secker-Walker LM, Chessells JM, Stewart EL et al. Chromosomes and other prognostic factors in acute lymphoblastic leukaemia: a long-term follow-up. Br J Haematol 1989; 72:336-342.
24. Bloomfield CD, Goldman AI, Alimena G et al. Chromosomal abnormalities identify high-risk and low-risk patients with acute lymphoblastic leukemia. Blood 1986; 67:415-420.
25. Secker-Walker LM. Prognostic and biological importance of chromosome findings in acute lymphoblastic leukemia. 10th anniversary article. Cancer Genet Cytogenet 1990; 49:1-13.
26. Rowley JD. Human oncogene locations and chromosome aberrations. Nature 1983; 301:290-291.

27. Erikson J, Nishikura K, ar-Rushdi A et al. Translocation of an immunoglobulin kappa locus to a region 3' of an unrearranged *c-myc* oncogene enhances *c-myc* transcription. Proc Natl Acad Sci USA 1983; 80:7581.
28. Erikson J, ar Rushdi A, Drwinga HL et al. Transcriptional activation of the translocated *c-myc* oncogene in Burkitt lymphoma. Proc Natl Acad Sci USA 1983; 80:820-824.
29. Klein G. The role of gene dosage and genetic transpositions in carcinogenesis. Nature 1981; 294:313-318.
30. Swansbury GJ, Secker-Walker LM, Lawler SD et al. Chromosomal findings in acute lymphoblastic leukaemia of childhood: An independent prognostic factor. Lancet 1981; ii:249-250.
31. Secker-Walker LM, Swansbury GJ, Hardisty RM et al. Cytogenetics of acute lymphoblastic leukaemia in children as a factor in the prediction of long-term survival. Br J Haematol 1982; 52:389-399.
32. Williams DL, Look AT, Melvin SL et al. New chromosomal translocations correlate with specific immunophenotypes of childhood acute lymphoblastic leukemia. Cell 1984; 36:101-109.

CHAPTER 2

Cytogenetic Techniques, Terminology and Classification In ALL

CYTOGENETIC INVESTIGATION

The purpose of a cytogenetic investigation of a patient with leukemia is to discover the karyotype (the chromosomal makeup) acquired by the cells of the malignant clone. The karyotype is described according to an internationally agreed system of cytogenetic nomenclature. The clonal karyotype is used to assign the patient to a chromosomal subgroup. In this chapter, I describe the salient features of chromosome preparation, problems and pitfalls in analysis, some basic guidelines to nomenclature and methods of classification in ALL. A glossary of cytogenetic terminology is given in Appendix 1.

The Outcome of Chromosome Investigations

A successful cytogenetic investigation of a patient with ALL is one in which an abnormal clone is detected or in which sufficient normal cells are available for analysis to rule out the presence of a clone. It is apparent from a number of surveys that >80% of successful analyses should reveal a chromosomally abnormal clone. The remaining 20% of cases reveal only normal cells and it is probable that in these cases leukemogenesis is associated with subchromosomal genetic change. Failure of cytogenetic analysis is, regrettably, a feature of this type of leukemia. Reported failure rates account for anything from 8% in single centers to 45% of cases in multicenter studies.[1-5]

Cytogenetic Techniques

The success or failure of a cytogenetic investigation depends to a considerable extent on the quality of the sample obtained and on the techniques used. Techniques for chromosome preparation will not be covered here as they are well described elsewhere.[6,7] It is, however, helpful to consider what we expect from a cytogenetic preparation and the steps we take to optimize the results.

An Adequate Sample

The first requirement is an adequate concentration of leukemic blasts. This is best achieved by a sample of bone marrow, where blood cells are formed. A peripheral blood sample is an acceptable alternative only if at least 30% and preferably 50% of blasts are present in the circulation. Part of the bone marrow or blood sample, taken from the patient before the start of treatment in order to diagnose the leukemia, is delivered to the laboratory for cytogenetic analysis. Collection of the sample into culture medium and rapid speed of transport to the laboratory are important to maintain the cells in a healthy viable state.

Yield of Metaphase Cells

Chromosomes in the cell can be recognized down the microscope only when they are in the condensed form they adopt during metaphase (the mid-point of cell division). Normally less than 3% of cells are dividing at any one time. It follows therefore that the leukemic cells which are investigated cytogenetically are highly selected and are restricted not only to the cells capable of undergoing cell division but also the small percentage of cells actually dividing at the time. The following steps are taken to increase the number of dividing cells. A Coulter counter is used to count the number of cells in the bone marrow sample. The correct number of cells for optimal growth (10^6/ml) is added to the culture flask containing culture medium and fetal calf serum. The culture flasks are put in the incubator at 37°C for 2, 24 or 48 hours to allow the cells to grow. To stop the cells dividing at metaphase, colcemid—a plant extract—which acts as a spindle poison and arrests the cells at metaphase, is added to the culture medium.

Separation of Chromosomes on the Slide

Chromosomes in the cell as well as cells on the slide can only be seen clearly if they are well spread out. To disentangle the chromosomes from each other, the cells in the culture medium are spun down in a centrifuge, the culture medium is removed and a hypotonic solution of potassium chloride (KCl) is added to them. The KCl enters the cell by osmosis which causes the cell membrane to expand and allows the chromosomes to disperse. The chromosomes are fixed with Carnoy's fixative—a mixture of three parts alcohol to one part acetic acid. To further separate both the cells and the chromosomes, these are dropped, suspended in fixative, onto clean, wet, cold slides. The slide is then passed through a spirit flame. As the alcohol ignites, the chromosomes are further dispersed before becoming firmly attached to the slide.

Band Staining

The staining technique produces stripes on the chromosomes and a banding pattern which is unique for each chromosome pair. The patterns reflect the uneven distribution of genes along the lengths of the chromosomes in which the gene-rich regions can be preferentially digested by an enzyme, trypsin. After trypsin digestion the chromosomes are stained with Giemsa. The gene-rich regions fail to take up the Giemsa stain and the characteristic alternating dark and light bands are produced on each chromosome.

Chromosomal Analysis and Detection of a Clone

For many years cytogenetic investigation of patients with ALL proved particularly difficult. The techniques in use today have been developed over many years. It is now clear that attention to detail at each stage in the procedure is necessary to optimize the results.[8-11] Of the utmost importance is analysis by trained personnel and the preparation of karyograms (the pictorial pairing of homologous chromosomes) to decipher rearrangements. There are now a number of image recognition systems on the market which greatly facilitate this task.

Clonal abnormalities are inferred when at least two cells with the same extra chromosome or structural rearrangement or three cells with the same missing chromosome are detected. The maximum incidence of chromosomally abnormal cases reported in ALL is >90%. Lower figures found in many studies most likely result from technical or procedural difficulties which are commonly encountered in studying this type of leukemia. During analysis every attempt must be made to find and analyze a chromosomally abnormal clone. The majority of patients with ALL have a bimodal population of cells in the bone marrow or peripheral blood at diagnosis: those with clonal chromosomal abnormalities and those which are chromosomally normal. The most likely explanation for these "normal" cells is that they belong to cell lineages not involved in the leukemia or are nonmalignant cells of the same cell type as those involved in leukemic process. For this reason, the coexistence of a chromosomally normal cell line is generally regarded as unimportant and may not always be reported. The alternative possibility, that chromosomally normal cells are malignant cells with submicroscopic genetic abnormalities, is probably true for the minority of cases in which no abnormal clone is found.

Detecting the Clone

What is clear is that the presence of a chromosomally normal population in the marrow may make it more difficult to discover the presence of an abnormal clone. For example, leukemic blasts have longer cell cycle times than their normal counterparts. The generation time of the blasts in ALL is about three days with DNA synthesis time of 20 to 25 hours, which are both significantly longer than those for normal cells. We know that the proliferative capacity of the leukemic clone shows considerable variation from patient to patient and at various stages of the disease. When the patient goes into remission the leukemic clone is not seen but in many cases the nonproliferating fraction retains the capacity to return to a cycling state as evidenced by return of the abnormal clone in relapse.[12,13] This means that normal cells in the marrow after 24 hours in culture may be overrepresented compared with leukemic cells. A further complication is added by the morphology of the chromosomes in malignant cells which is frequently less good than of the residual normal cells. The abnormal cells may be passed over as unanalyzable. Experience has shown it is imperative to attempt to analyze these poorly defined chromosomes.

Another consideration relates to the fact that two marrow aspirates or different portions of the same marrow aspirate from the same patient may contain different numbers of leukemic cells. Comparison of analysis in two different laboratories taking different portions of the same marrow aspirate from 73 children with ALL showed a similar rate of detection of clonal abnormalities (72%) in both laboratories. This suggested that overnight transport of the sample (which was necessary for one of the laboratories) did not compromise the results. However, the findings based on individual patient samples differed in some cases, with one laboratory obtaining a result or detecting a clone and the other not. This suggested either that the sample had been unequally divided or that certain clones survived transportation better than others. When the data from the two laboratories were combined the overall detection rate of an abnormal clone was 87%.[14]

Chromosomal Analysis and a Classification of Normal

The incidence of patients with ALL without clonal abnormalities in the leukemic cells varies widely between series. Some of these differences no doubt relate to variations in technique but it is clear that at least some of these patients have submicroscopic genetic change in their malignant cells which does not result in a rearrangement of their chromosome number or structure. In other cases, it is entirely possible that by chance, the dividing cells and those chosen for analysis were not part of the malignant clone. In cases classified as normal, there may be single cells with chromosomal abnormalities. They are referred to as random abnormalities because there is no evidence that these belong to a clone.

Random Abnormalities

Random chromosomal abnormalities are of two kinds. Chromosome gaps or breaks are known to be unstable configurations which may be induced in a random fashion by viruses, environmental carcinogens or by treatment. This is the only kind of change that can truly be excluded as part of the malignant clone. Cells with abnormalities of chromosome number or structure are a feature of malignant cell populations. When these are found as single cells their importance is difficult to assess. Single abnormal cells may result from technical artifact such as those with chromosome loss resulting from the rupture of fragile cells during chromosome preparation. Certain types of chromosomal abnormality seen as random change may be incompatible with the survival of the cell, making it incapable of forming a clone. However, the presence of a single cell with one of the abnormalities known to occur in malignant clones may well represent a minor or relatively inactive clone. Cytogenetic investigation alone is unable to distinguish between these possibilities. Convention dictates that such single cells are presumed to be random events.

Exclusion of a Clone

A major consideration in classifying a patient "chromosomally normal" is the number of cells available for analysis. Before the mid 1980s it was accepted that patients with five fully analyzed cells should be classified as normal. This is now recognized as inadequate. The detection of a clone and degree of confidence with which the presence of a clone can be excluded is dependent on the number of cells analyzed. For example, analysis of 20 cells all found to be normal excludes the presence of an 11% actively dividing clone with 90% confidence. To exclude the presence of a 5% clone with 99% confidence requires analysis of 112 cells.[15] (see Table 2.1). Because chromosome analysis is a time consuming and labor intensive exercise, the number of cells analyzed has to be a compromise. The requirement for a classification of "normal" in current ALL treatment trials in the U.K. is full analysis of 20 cells, which excludes the presence of an 11% clone with 90% confidence.

When Chromosomal Analysis Fails

Cases in which metaphases are absent, insufficient or of inadequate morphology for the detection of a clone are classified as failures. Samples most likely to fail are peripheral blood, acellular marrows and samples which have spent more than 24 hours in transit.[10,11] There is, however, a hard core of cases in any series

Table 2.1. The number of cells required for analysis to exclude the presence of a clone in between 1% and 38% of cells with 90%-99% confidence.

No. cells analyzed	Confidence levels 0.90	0.95	0.99
5	38%		
10	21%	26%	37%
15	15%	19%	27%
20	11%	14%	21%
25	9%	12%	17%
30	8%	10%	15%
50-55	5%	6%	9%
99-112	3%	3%	5%
≥459	1%	1%	1%

Adapted from Hook E.B.[15]

which fail for no apparent reason suggesting that failure of cell division in vitro is inherent in a proportion of leukemic bone marrows.

NOMENCLATURE

The karyotype, (chromosomal makeup of a cell) is described according to convention, currently ISCN 1995.[16] The following description is intended as a guide to this terminology rather than a comprehensive description of it (see Symbols and Abbreviations and Appendix 1).

The Normal Human Karyotype

Normal human somatic cells contain a diploid set of 46 chromosomes in 23 pairs made up of 22 pairs of autosomes numbered 1-22 and a pair of sex chromosomes XX (female) or XY (male). Chromosomes are derived equally from the haploid set of 23 chromosomes in each parental germ cell.

A diagrammatic representation (ideogram) of the 24 chromosomes found in humans is shown in Figure 2.1. Each chromosome has a long (or q) arm and a short (or p) arm separated by a primary constriction, the centromere. The chromosome arms are divided into regions and subregions based on the bands and

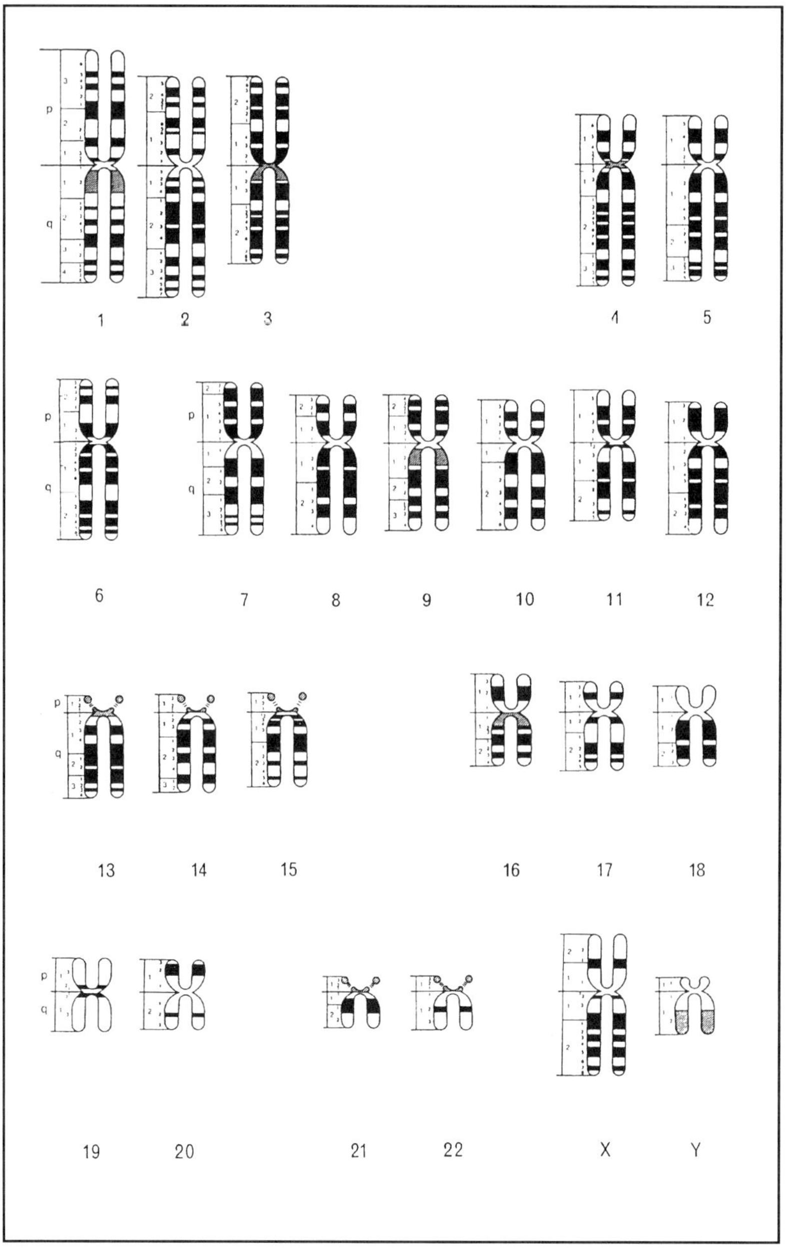

Fig. 2.1. Ideogram of a male human karyogram with approximately 300 chromosome bands as seen with trypsin-Giemsa staining (G-bands). Adapted from ISCN 1991.[17]

numbered from the centromere out towards the ends (telomeres). This enables abnormalities to be described in terms of the affected region of the chromosome. Thus 11q23.2 refers to the long arm of chromosome 11, region 2, subregion 3, subsubregion 2. In hematological disorders, chromosomes are seldom sufficiently extended to identify all but a few subsubregions.

CHROMOSOMAL ABNORMALITIES IN ALL

Clonal abnormalities found in ALL may be numerical changes, structural changes or a combination of the two.

NUMERICAL CHANGE

Numerical change involves either chromosome gain resulting in three, four, etc. instead of two copies of one pair of homologs resulting in trisomy, tetrasomy, etc. for that chromosome (indicated by the prefix +). The loss of a chromosome results in monosomy for that chromosome (indicated by the prefix -).

STRUCTURAL CHANGE

Structural change includes deletions (del) when part of one of the chromosome's arms is missing, e.g., del(11)(q23) is loss of material distal to 11q23, or translocations (t) when chromosomal material is repositioned on another chromosome. Balanced translocations are assumed to be reciprocal; they usually result in shortening of one chromosome and lengthening of the other. The terms p- or q- and p+ or q+ (with the - or + as a suffix) are loosely used to describe the resulting chromosome made up of two or more chromosomes referred to as derived chromosomes. An ideogram of a balanced translocation is shown in Figure 2.2. A reciprocal translocation may be between three or more chromosomes (complex). Inversion (inv) describes the inversion of the segment of the chromosome lying between two bands.

An abnormal clone may have numerical change only, structural change only or a combination of the two. Any clone with numerical change shows a change of ploidy from the normal diploid (46 chromosomes) complement and is classified by the number of chromosomes in the clonal cells as less than (hypo) or greater than (hyper) the nearest ploidy number as follows: near haploid

23 to 34 chromosomes; hypodiploid 35-45 chromosomes; hyperdiploid 47-57 chromosomes; hypo or hyper triploid (69 ± range 58 to 80 chromosomes); near tetraploid (92 range 81-103 chromosomes). A clone with 46 chromosomes with structural change such as deletions or translocations is pseudodiploid.[17] Examples of chromosomal change and their genetic consequences are summarized in Figure 2.3.

An example of the karyotype of a cell with structural and numerical change written in ISCN 1991 is shown here. The number of chromosomes in the cell is given first, followed by the sex chromosomes, followed by a description of the abnormal chromosomes in numerical order. The number of cells with that karyotype then follows in square brackets. Thus 50,XY,+7,+8,t(9;22)(q34;q11),+13,+19[18]/46,XY[2] describes a dominant clone of cells which are hyperdiploid (having 50 chromosomes) due to the gain of chromosomes 7,8,13 and 19. There is also structural change, namely a balanced translocation between chromosomes 9 and 22 with breaks at 9q34 and 22q11.

CHROMOSOMAL CLASSIFICATION

Patients with ALL are classified by the chromosomal findings in the leukemic cell population. Patients with an abnormal clone may be classified by ploidy, by structural change or both.

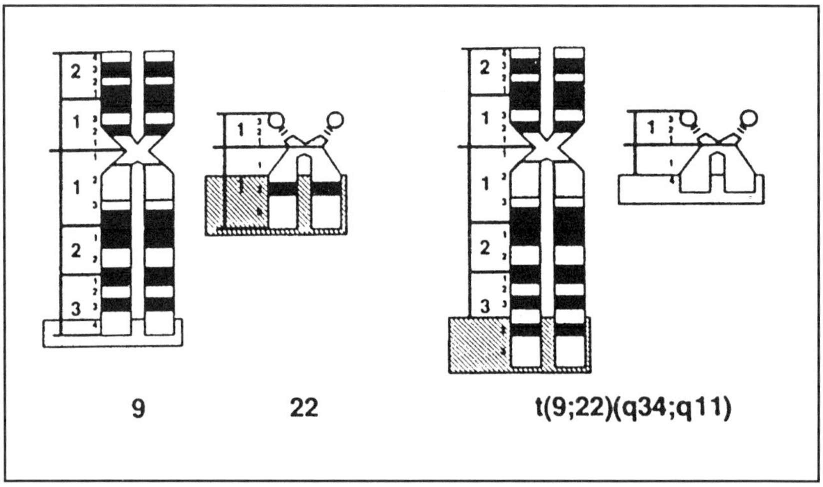

Fig. 2.2. Ideogram of translocation t(9;22)(q34;q11). From Cancer Cytogenetics 2nd Edition. Heim S, Mitelman F. Copyright © 1995. Reprinted by permission of John Wiley and Sons, Inc.

Established Chromosomal Change, Subgroups and Classification

Established or recurring primary abnormalities are those which have been detected as the only change in at least five individual cases of ALL from at least two different laboratories worldwide.[18] These include structural abnormalities and numerical change involving the loss (monosomy) or gain (trisomy) of a single chromosome. Those identified at the Human Gene Mapping conference in 1991 (HGM 11) are listed in Table 2.2.[18] Many of these are also found in association with other chromosomal changes. Equally important in ALL are clones showing a gain of several chromosomes together to result in hyperdiploidy with 47-50 chromosomes (low hyperdiploidy) or with >50 chromosomes (high hyperdiploidy. Hyperdiploid clones may also contain structural change.

Established structural changes listed in Table 2.2 are of two main kinds. Some of these are associated with specific clinical features and

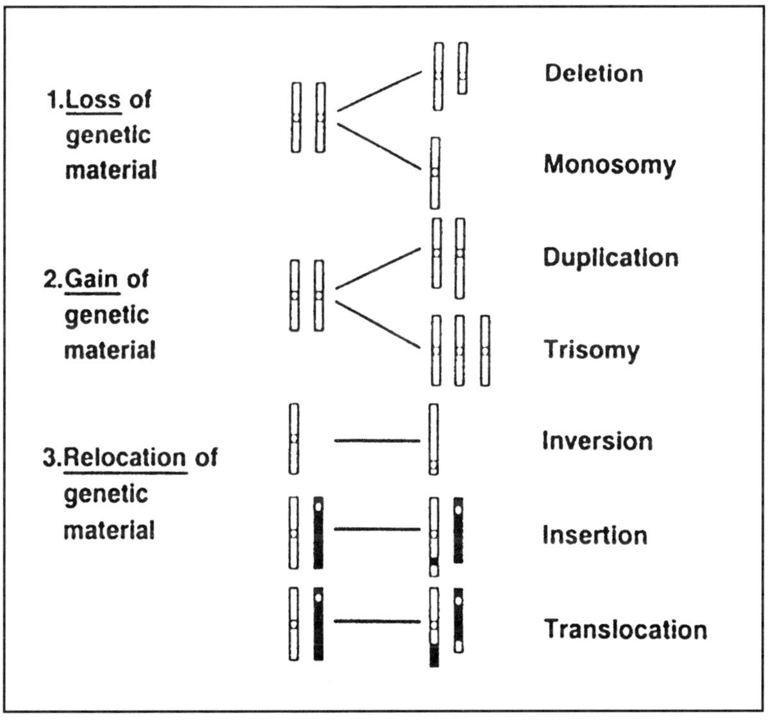

Fig. 2.3. Methods of gain, loss and rearrangement of genetic material through chromosomal rearrangements. From Cancer Cytogenetics 2nd Edition. Heim S, Mitelman F. Copyright © 1995. Reprinted by permission of John Wiley and Sons Inc.

Table 2.2. Primary nonrandom abnormalities in ALL established at HGM 11

>10 cases	**<5 cases**
t(1;19)(q23;p13)	t(5;14)(q31;q32)
t(4;11)(q21;q23)	dic(7;9)(p13;p11)
del(6)(q13-24;q21-27)	del(7)(q11)
+8	del(7)(q32)
del(9)(p13-22)	
t(9;22)(q34;q11)	t(8;12)(p12-22;q13)
t(11;19)(q23;p13)	t(8;14)(q11;q32)
-20	dic(9;12)(p11-13;p11-12)
+21	t(9;12)(p11;p12)
del(22)(q11-q13)	i(9q)
B-ALL	del(11)(q14-q23)
t(8;14)(q24;q32)	t(12;17)(p12-13;q12)
T ALL	+13
t(10;14)(q24;q11)	del(14)(q11-24-q22-32)
t(11;14)(p13;q11)	t(14;22)(q32;q11)
	+16
5-10 cases	+18
	-21
t(1;11)(p32;q23)	-X
t(9;11)(p21-22;q23)	B-ALL
	t(2;8)(p12;q24)
del(12)(p11-13)	T ALL
del(13)(q11-22-q14-34)	t(1;14)(p32-34;q11)
-Y	t(7;11)(q35;p13)
B-ALL	t(7;14)(q35-36;q11)
t(8;22)(q24;q11)	t(8;14)(q24;q11)

Adapted from Mitelman et al.[18]

have blasts which express a particular immunophenotype. These include the translocations found in mature B cell and T ALL which involve the chromosomal location of one or more of the immunoglobulin or T cell receptor genes, respectively. Other abnormalities are associated with a diversity of clinical or blast cell features. Each of these are important both for their prognostic importance and for the role they have played in unravelling the mechanisms by which chromosomal change is implicated in leukemogenesis.

CLASSIFICATION AND PROGNOSIS

Chromosomal classification of ALL has undergone several changes since classification by ploidy (number of chromosomes in the clone) was first shown to be of prognostic importance in 1978.[19] The Third International Workshop on chromosomes in leukemia investigated the prognostic importance of four different methods of classification.[20] Comparison of abnormal and normal cases was found to be helpful only in the adult series but the presence of a normal cell line in patients having an abnormal clone was without effect. Prognostic significance was demonstrated for the other three methods; the presence or absence of a translocation, classification by ploidy alone and a hierarchical classification (Lund) by which patients were classified first by the presence of one of four structural changes and the remainder were classified by ploidy.[2,20,21] In other studies, patients have been classified twice, both by structural change and also by ploidy. All methods of classification have been shown to be prognostically important and to add prognostic significance to the other variables, such as age, white blood cell count, FAB type and sex. It has recently been demonstrated that for adults a hierarchical classification, in which structural change takes precedence over ploidy, is most valuable in terms of prognosis.[22]

CHROMOSOMAL ABNORMALITIES DETECTED BY MOLECULAR TECHNIQUES

Many of the genetic changes brought about by chromosomal change can now be identified by molecular techniques. These have extended the scope of cytogenetic analysis by enabling the presence of a chromosomal change to be detected in whole cell populations and in resting as well as in cycling cells (see chapter 9).

CONCLUSION

Cytogenetic and molecular techniques used together now provide a powerful combination for the classification and care of patients with ALL. The details of these approaches will be described in the rest of the book.

CHAPTER 3

Modal Number and Ploidy Classification

NUMERICAL CHANGE AND PLOIDY CLASSIFICATION

The modal number is the commonest number of chromosomes found in the cells of a malignant population. The ploidy of the clone is decided by the modal number as defined by ISCN 1991.[1] Ploidy may be near haploid (23), near diploid (46), near triploid (69) or near tetraploid (92).

Clones with numerical change have too many or too few copies of one or more chromosomes. These can arise either from an unequal distribution of chromosomes to the daughter cells at cell division due to lagging and nondisjunction of one or more chromosomes or from selective endoreduplication of certain chromosome(s). Cells with chromosome loss (monosomy of one or more chromosomes) are referred to as hypodiploid and cells with chromosome gain (trisomy of one or more chromosomes) are hyperdiploid. Pseudodiploid refers both to clones with chromosome loss matched by chromosome gain and to clones with 46 chromosomes in which one or more chromosomes have undergone structural change. In acute lymphoblastic leukemia, only certain chromosomes contribute to monosomy or trisomy. It therefore appears that only certain chromosome configurations are compatible with the viability of the cell and the ability to form clones. Chromosomally abnormal cells are frequently accompanied by a chromosomally normal cell line. The detection of normal cells is partly a feature of the number of cells analyzed—with a sufficiently large sample these would possibly be found in all cases.

with ALL. These were 14 boys and 17 girls, aged nine months to 17 years. Prognosis reported for this group has been variable. It has been associated with an excellent prognosis,[4,10] with a prognosis no worse than any other ploidy group except hyperdiploid >50[5] and with a poor prognosis.[3,11-13] These conflicting outcomes suggest that prognosis may be decided by particular chromosomal abnormalities within the clone rather than by chromosomal loss per se.

Hypodiploidy can be further subdivided into three subgroups according to the numbers of chromosomes in the clone. These are 40-45 chromosomes, 30-39 chromosomes and <30 chromosomes (near haploidy).[5,14]

Clones with 40-45 Chromosomes

40-45 chromosomes

In a study by Pui et al, 26/31 or 84% hypodiploid cases had a clone with 45 chromosomes. These were indistinguishable by cellular DNA content from patients with normal diploidy. Chromosome 20 was lost most frequently, and in three cases monosomy 20 was the only change. Cases with 45 chromosomes included three patients with dic(9;12)(p11;p12) (see also chapter 6). Structural change was seen in 22/26 cases with 45 chromosomes. A single case was found with 43 chromosomes and complex change.[5]

Monosomy 7

Monosomy 7 is a rare event in ALL. It has been described as the only change in less than half a dozen cases.[15-17] Although there

Table 3.2 Distribution between ploidy groups of adults with ALL

Total Cases	N %	Ho %	PS %	HeL %	HeH %	T/T	
73	38.5	9.6	32.9	13.7	5.5	0	Walters et al[49]
350	33.7	6.9	35.4	10.3	10.0	3.7	Secker-Walker et al[28]
156	19.0	7.7	47.0	8.3	13.5	5.0	Appendix 2

are examples of monosomy 7 in CD10+ cases there is some evidence that the target cell for monosomy 7 in ALL might be a multipotent stem cell, capable of giving rise to cells of the myeloid as well as the lymphoid lineage.[16]

Monosomy 20

Monosomy 20 has been described in 4/168 (2.4%) and in 1/90 cases of childhood ALL. At least 12 cases have been described altogether. All these have been children, mainly with B lineage but also with T ALL. Survival appears to have been good but followup of the reported cases has been relatively short for childhood ALL.[18,19]

Clones with 30-39 Chromosomes

Also referred to as severe hypodiploidy, the group with 30-39 chromosomes is extremely rare, particularly in childhood where it accounts for <0.3% of cases.[5] In a study of 22 hypodiploid cases, Callen et al noted that among 11 patients with 30-39 chromosomes five were aged over 40 years and only three were children. The presence of a second, hyperdiploid cell line, with a doubling of the hypodiploid number was commonly seen. It was observed that many of these patients had both small and large blasts with numerous small vacuoles. These authors suggested that clones with 30-39 chromosomes may be indicative of secondary or environmentally induced ALL.[14]

Near Haploidy <30 Chromosomes

Near haploidy is marked by having both sex chromosomes XX or XY and the gain from haploidy (23 or 24 chromosomes) of three or four chromosomes with a preferential gain of numbers 10, 14, 18 and 21. An example of a near haploid metaphase spread is shown in Figure 3.2. The reported incidence ranges from 0.7% to 2.4%.[5,20] The oldest reported patient with near haploidy was 19 years old. Among the 28 reported cases there have been more females than males (16:12). A median WBC 14 x 10^9/L (range 1.0 -354) and B-precursor ALL further characterize the group.[14,20] In general, prognosis for patients with near haploidy has been poor with a median survival of 11 months. Two long term survivors have been reported: one in first remission at seven years and the other in second remission at 10 years.

Near Haploidy and Hyperdiploidy

There have been several reports of patients with near haploidy having a second clone with hyperdiploidy that represents an exact or near doubling of the chromosomes in the near haploid cell line. A karyogram of a clone which may have arisen in this way is shown in Figure 3.3.[14,20-23] Whether in these cases prognosis relates to the near haploidy (poor) or to the hyperdiploidy (good) is not known. There is anecdotal evidence that such cases may have a good prognosis. One such case had a remission duration of at least four years.[5]

PSEUDODIPLOIDY

Cases with a chromosome number of 46 but with an abnormal chromosomal configuration are pseudodiploid. In all series, this is the largest group and both chromosomally and clinically it is always heterogeneous. Rarely a pseudodiploid clone may be formed by the gain of one or more chromosomes matched by chromosome loss, but the impact of this type of change has not yet been explored. Over 90% of pseudodiploid clones have structural

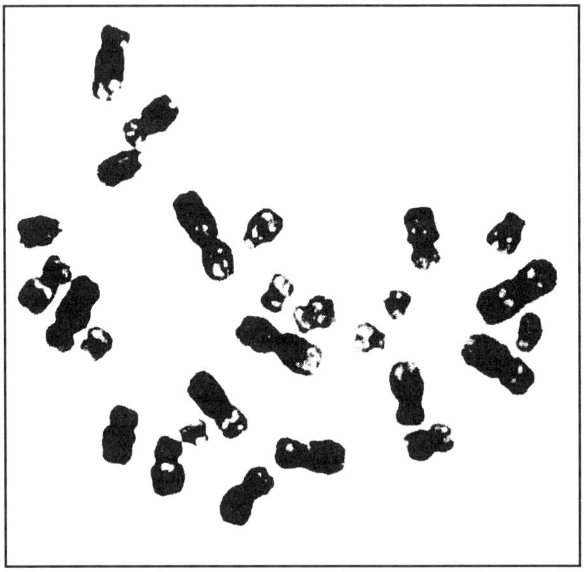

Fig. 3.2. Near haploid metaphase spread with 27,XY,+6,+10,+21 from a child with ALL. Reproduced with permission from Gibbons et al.[20]

change and this may involve any or several chromosomes and it may be simple, involving a single translocation or partial arm deletion or complex, involving more than one change. The known poor-risk translocations t(4;11) (see chapter 4) and t(9;22), (see chapter 5) most frequently occur in pseudodiploid clones. These two translocations together contribute significantly to pseudodiploidy in adults since they account between them for between 16% and 30% of patients. In childhood (patients aged between one and 14 years), they account for 7% together.

Clinical Features

The reported incidence of pseudodiploidy in childhood series ranges from 25% to 40% (mean 29%) see Table 3.1.[10,13,24-27] Surveys with a low incidence of cases described as normal (with no abnormality detected) have correspondingly higher percentages of both pseudodiploid and high hyperdiploid patients. The incidence of pseudodiploidy is higher in adults than in children (see Table 3.2). Of 350 adults in the MRC trial UKALLXA, 124 (35%)

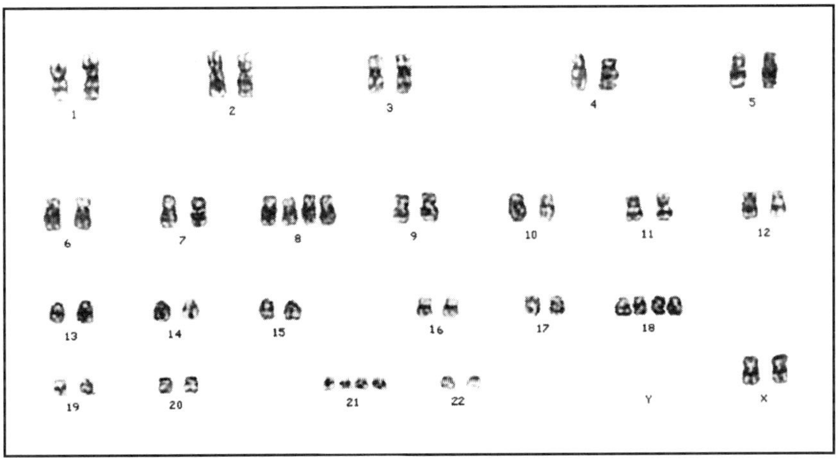

Fig. 3.3. Karyogram of a high hyperdiploid clone 52,XX,+8,+8,+18,+18,+21,+21, which may have arisen from a near haploid cell line.

were pseudodiploid. Of these, 87 (70%) had established structural change. When adults were classified hierarchically (established structural change taking precedence over ploidy) this left only 37/350 (11%) of cases in the pseudodiploid category.[28] Pseudodiploidy in all series, childhood and adult, has been associated with poor-risk features of high WBC and non C-ALL.[29]

PROGNOSIS

It is important to distinguish the prognosis for pseudodiploid patients classified by ploidy alone from those classified hierarchically. When patients are classified by ploidy only, pseudodiploidy has a poor prognosis. In recent childhood studies, event-free survival has been poorest for pseudodiploid patients with event-free survival at between four and six years ranging from 22%, in a study of Japanese children,[26] to 57% in France,[27] to 63% at St Jude's (see Table 3.3).[13,29] In adults on UKALLXA, event-free survival of the pseudodiploid patients was not significantly different from all other patients combined whether classification was by ploidy alone or hierarchical. Using the hierarchical grouping, 25% of patients were event-free at three years compared with 36% of all other patients.[28]

CHROMOSOMALLY NORMAL

The percentage of patients in whom no clone is detected is dependent both on techniques and on the number of cells analyzed to exclude the presence of a clone. Childhood studies have reported as normal up to 40% of cases,[3,30] but more recent studies show that normal may be as few as 9%[13] or as high as 37%.[25] (mean 26%) (see Table 3.1). It is still a matter of debate how many cells should be analyzed to exclude the presence of a clone before a classification of normal should be made. Curently in the U.K. a minimum of 20 cells are required and here, 20% of all patients have been classified as normal (see Appendix 2). St Jude's Children's Hospital in Memphis, Tennessee, U.S.A. has consistently reported a very low incidence of normals as shown recently by Raimondi et al.[13] A minimum of five cells analyzed was accepted at the Third Workshop and for UKALLXA and these studies showed 31% and 34% normals respectively in adults.[3,28]

Table 3.3 The percentage of patients surviving at 4 or 5 years by ploidy group in childhood ALL

	Van der Plas et al[10] SR	Kobayashi et al[26]	GFCH[27]	Raimondi et al[13]
Time from Diagnosis	5 yrs	4 yrs	4 yrs	4 yrs
Ploidy Group				
HeHN	92%	74%	85%	84%
HeH+S	73%			
HeL	88%	68%	67%	77%
PS	64%	22%	57%	63%
N	70%	55%	63%	87%
HO	100%	54%	71%	46%

N = numerical change only
+ S = with structural change
SR = standard risk cases

Clinical Features

Raimondi et al have noted an excess of normals among patients with T ALL. T ALLs having up to 30% normal cases compared with 10% of normals among patients with other immunophenotypes at the same center.[13,31] Of 59 adults with T ALL on UKALLXA, 25 (42%) were chromosomally normal.[28]

Prognosis

In most surveys of childhood ALL, a classification of normal has shown an intermediate prognosis (see Table 3.3). Adults in the normal subgroup were shown to have a relatively good prognosis at the Third Workshop when first examined and after a minimum of six year followup.[3] This was not seen using a hierarchical system in UKALLXA when the five year projected disease-free survival for the normal category was among the poorest at 26% among subgroups in which prognosis ranged from 69% to 14%.[28]

HYPERDIPLOIDY

Clones with >46 chromosomes are hyperdiploid—a finding first shown to be associated with other good-risk factors in 1978. Patients with a hyperdiploid clone were most likely to be aged two to nine years, of female sex, with a leukocyte count <10 x 10^9/L and with C-ALL immunophenotype.[2,4,32] At the Third Workshop, the hyperdiploid category was subdivided on the basis of a less favorable group with low hyperdiploidy with 47-50 chromosomes and the most favorable group with high hyperdiploidy with >50 chromosomes. The upper limit in this category has been variously set at 60, 65 or 68 chromosomes. Two more hyperdiploid categories have been identified: near-triploidy with between 66 and 80 chromosomes and near tetraploidy with >80 chromosomes.[30,33]

LOW HYPERDIPLOIDY 47-49/50 CHROMOSOMES

Raimondi et al have described 86 children with 47-50 chromosomes. Numerical change only was seen in 21 cases (24%) and the remaining cases also had structural change (76%). Most cases had clones with 47 chromosomes (72%) followed by 48 chromosomes (20%), 49 chromosomes (5%) and 50 chromosomes (3%) The most frequent gains were of chromosomes 21, X, 8 and 10, and structural change included abnormalities of 1q, 6q, 12p and 19p.[34] Moorman et al[35] showed that chromosomes gained more frequently than expected in low hyperdiploid clones were numbers 21 (42%), X (22%), 10 (13%), 8 (11%), 6 (9%) and 16 (7%).

CLINICAL FEATURES

The incidence of low hyperdiploidy in recent reports of childhood ALL ranges from 7% to 13% (mean 11%) (see Table 3.1)[10,13,24-27] and is similar in adults (see Table 3.2). There is little consistency in the clinical and blast cell findings of this group which reflects the heterogeneous nature of chromosomal change in this group.

PROGNOSIS

Prognosis in different series also reflects the heterogeneous nature of this subgroup. There has been no evidence that structural change adversely affects prognosis in these patients. Prognosis for this group appears to have been improved by more recent treatment protocols.[34] Patients entered to the Third Workshop had a relatively poor prognosis on short-term followup[3] but with a minimum followup of six years, the low hyperdiploid cases fared almost as well as those with high hyperdiploidy.[12] Event-free survival for this group of adults in UKALLXA at three years was 53% compared with 59% for high hyperdiploidy.[28]

Trisomy 4

Trisomy 4 has been described as the only change in two patients with ALL. One of these was a 42 year old Japanese man with T ALL,[36] The other was a 27 year old French woman with early B ALL (CD19+ve, CD10-ve).[37] Further cases will be needed to define the target cell in such cases and to determine the clinical outcome in patients presenting with trisomy 4 in ALL.

Trisomy 5

Trisomy 5 as the sole abnormality has been described in a 26 year old male with C-ALL who was refractory to treatment[38] and in a 24 year old woman with T ALL who survived for at least one year from diagnosis.[39] More cases are needed to verify the target cell for chromosomal change and the clinical relevance of trisomy 5.

Trisomy 8

The incidence of trisomy 8 as the only change in ALL is similar in childhood (1.7%) and in adults (1.3%). In a review of seven published cases it was shown to have occurred in B-lineage and in T ALL in children aged from nine months to 13 years and in two adults aged 36 and 39 years.[40] A more recent survey from the Pediatric Oncology group identified trisomy 8 as the only change in 7/2690 or 0.3% of children with ALL.[41] The prognostic implications of this await clarification but it appears that trisomy 8 alone and without clonal evolution may have a relatively good outcome.[40-42] A recent review suggested that trisomy 8, reported in some hematological malignancies, may represent undiagnosed constitutional trisomy 8 mosaicism and this deserves further investigation.[43]

Trisomy 21

Trisomy 21 as the sole acquired karyotypic change has been reported by the St Jude's group in 21/601 (3.5%) of childhood ALL. None of these patients had evidence of constitutional trisomy 21. Patients were aged one to 18 years (median 3.3 years) and had WBCs ranging from 1.8 to 82 (median 11.6) x 10^9/L. All patients had B lineage ALL, either C-ALL or preB. Followup ranged from one month to 6.4 years. Two patients had relapsed, one in the testis and one in the central nervous system, but none of the patients had died at the time of the report. It was suggested that trisomy 21, as the only change, may be associated with a favorable prognosis.[44,45]

High Hyperdiploidy 50/51-60/65/68 Chromosomes

High hyperdiploidy includes a range of chromosome numbers between 50 and 68. The metaphases of a single case may exhibit a range of chromosome numbers but it is not generally clear whether these are indicative of clonal instability or whether chromosomal loss in some cells is a technical artifact. The clonal number may be given as the mode (most frequently seen), or may be given as a range. In some instances, clonal evolution is apparent. Chromosomal gain appears to be restricted to certain chromosomes. Moorman et al[35] have shown that, in 146 high hyperdiploid patients, chromosomes which were gained more frequently than expected were numbers 21 (100%), X (95%), 14 (90%), 6 (86%), 18 (78%), 4 (75%), 17 (65%) and 10 (61%). A karyogram showing typical chromosome gains in a high hyperdiploid clone is shown in Figure 3.4. Structural change as well as chromosome gain is found in at least 62% of patients and this includes the poor-risk translocations t(4;11), t(8;14) and t(9;22). For example amongst 76 children with Ph+ALL, 10% had a high hyperdiploid clone. Among 54 Ph+ adults, 20% had a high hyperdiploid clone (see Table 5.1).[28,46-48]

Clinical features

In all surveys reported to date, patients with >50 chromosomes have had clinical features associated with good-risk ALL. Clinical findings associated with high hyperdiploidy include: age two to nine years or age <18 years in adult series, leukocyte counts <10 x 10^9/L, female sex, C-ALL immunophenotype and FAB type L1. The incidence in recent childhood series has ranged between 20% and 33% (mean 27%) but is considerably lower in adults (5% to 12%).[28,49,50]

Prognosis

Hyperdiploidy is associated with a better prognosis than that of any other ploidy group. Three childhood surveys have shown an average five year event-free survival of 72%.[10,51,52] In recent studies, event-free survivals of 74% to 88% have been seen at four and six years (see Table 3.3).[10,13,26,27,53] The benefit of hyperdiploidy >50 was emphasized in a recent study of childhood ALL with followup of 12-18 years. This showed karyotype and treatment alone to be of prognostic importance with overall survival in patients with

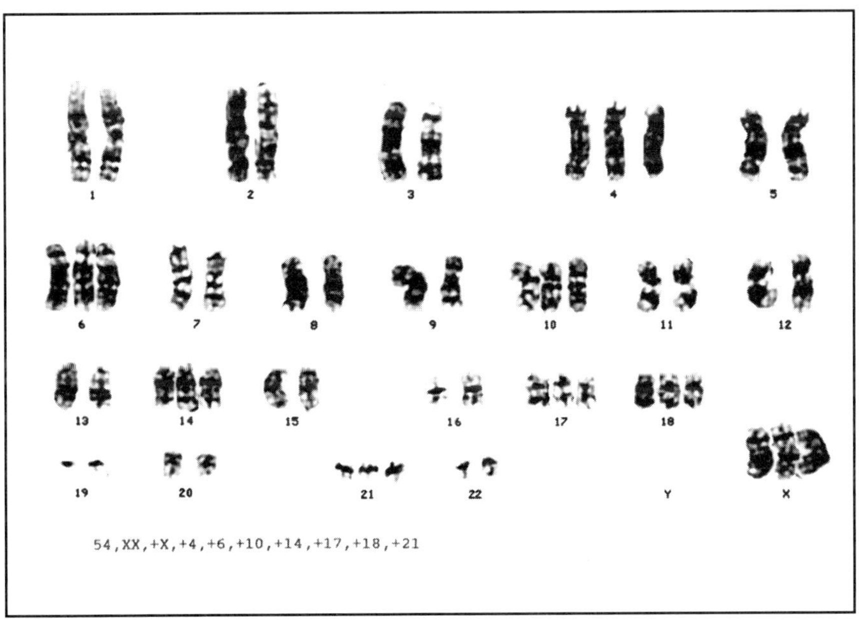

Fig. 3.4. Karyogram of a high hyperdiploid clone 54, XX+X, +4, +6, +10, +14, +17, +18, +21.

as five years from diagnosis.[54] On the other hand these patients may enjoy longer second remissions than those from other chromosomal subgroups (see Fig. 3.6).[59]

Drug sensitivity

The mechanism of cure in these patients is unknown. Clearly a clone which is eliminated must be sensitive to the treatment given. Treatment includes the folate antagonist, methotrexate (MTX), and it has been shown that children with B-lineage ALL, whose lymphoblasts accumulate high levels of methotrexate and methotrexate polyglutamates in vitro, experience better event-free survival than those whose lymphoblasts do not. Levels of methotrexate polyglutamates were shown to be significantly higher in children with >50 chromosomes than in those with pseudodiploidy or normal chromosomes p < 0.001).[60] In addition, hyperdiploid cells from 16 patients were shown to be significantly more sensitive to antimetabolites than the cells of 74 children with nonhyperdiploid cells in vitro. This was especially true for mercaptopurine and l-asparaginase.[61]

Survival of hyperdiploid clones in vitro

Manabe et al have reported prolonged survival of B-lineage leukemic cells on feeder layers of bone marrow-derived stromal cells in a serum-free environment. The only cells not giving rise to viable cultures were those from two patients with hyperdiploidy greater than 50 chromosomes. Thus hyperdiploid blasts were less likely to survive in vitro than those with other chromosomal abnormalities. This greater propensity to apoptosis than that of clones with other chromosomal configurations may help to explain the early and permanent response to treatment shown by the majority of patients with hyperdiploid clones.[62]

Mitotic index and early elimination of the hyperdiploid clone

The notion that sensitivity to drugs is a feature of high mitotic activity and that cure is associated with early elimination of the clone has been investigated by Heerema at al using chromosome-specific centromeric probes to investigate interphase nuclei in high hyperdiploid cases. Comparison of the size of the clone determined by cytogenetic analysis of metaphase cells and cells in

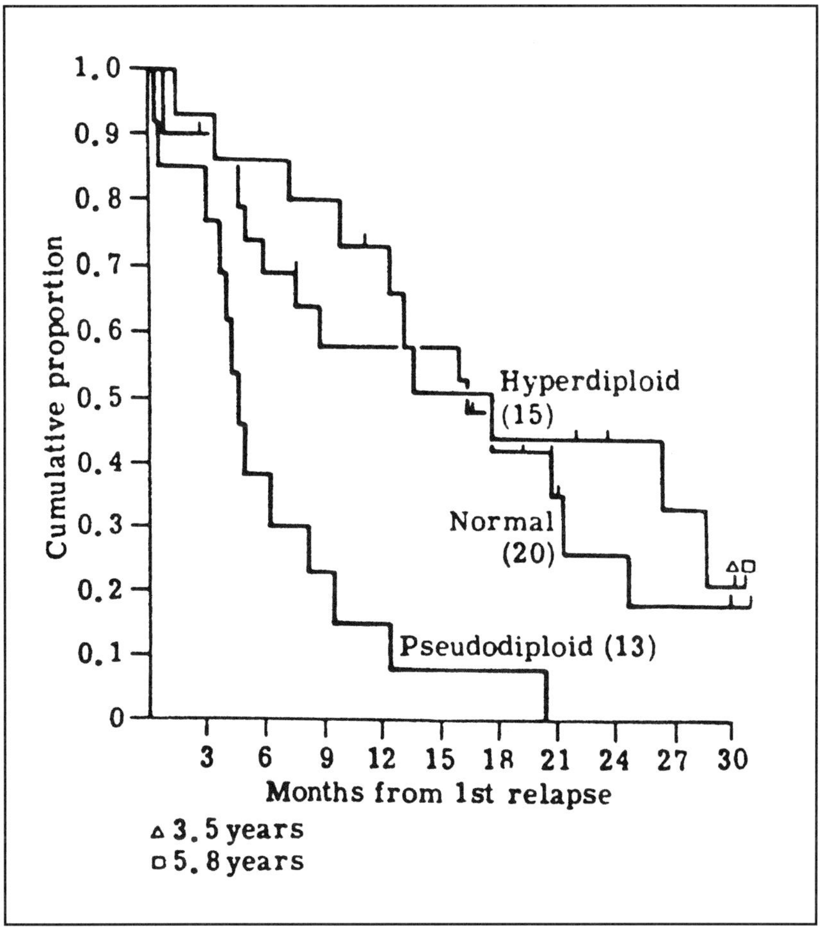

Fig. 3.6. Survival following first relapse by ploidy group. From Secker-Walker[59] copyright 1984 reprinted by permission of the publisher Elsevier Science, Inc.

interphase in a series of children exhibited heterogeneity of the mitotic activity of the clone at diagnosis. Samples examined in cytogenetically normal remission retained evidence of the hyperdiploid clone in 5/7 samples one month from diagnosis which had been eliminated by two months.[63] Larger studies with longer followup will be needed to determine the relationship between the retention of the clone in early hematological remission and outcome.

TRIPLOIDY AND TETRAPLOIDY >68 CHROMOSOMES

Cases with >65 chromosomes are rare in ALL at all ages but are probably more frequent in adults. An example of a near triploid

6. Prigogina EL, Puchkova GP, Mayakova SA. Nonrandom chromosomal abnormalities in acute lymphoblastic leukemia of childhood. Cancer Genet Cytogenet 1988; 32:183-203.
7. Heinonen K, Rautonen J, Siimes MA et al. Cytogenetic study of 105 children with acute lymphoblastic leukemia. Eur J Haematol 1988; 41:237-242.
8. Secker-Walker LM, Chessells JM, Stewart EL et al. Chromosomes and other prognostic factors in acute lymphoblastic leukaemia: a long-term follow-up. Br J Haematol 1989; 72:336-342.
9. Uckun FM, Gajl Peczalska KJ, Provisor AJ et al. Immunophenotype karyotype associations in human acute lymphoblastic leukemia. Blood 1989; 73:271-280.
10. Van der Plas DC, Hahlen K, Hagemeijer A. Prognostic significance of karyotype at diagnosis in childhood acute lymphoblastic leukemia. Leukemia 1992; 6:176-184.
11. Bloomfield CD, Goldman AI, Alimena G et al. Chromosomal abnormalities identify high-risk and low-risk patients with acute lymphoblastic leukemia. Blood 1986; 67:415-420.
12. Bloomfield CD, Secker-Walker LM, Goldman AI et al. Six-year follow-up of the clinical significance of karyotype in acute lymphoblastic leukemia. From the Sixth International Workshop on Chromosomes in Leukemia 1987. Cancer Genet Cytogenet 1989; 40:171-185.
13. Raimondi SC. Current status of cytogenetic research in childhood acute lymphoblastic leukemia. Blood 1993; 81:2237-2251.
14. Callen DF, Raphael K, Michael PM et al. Acute lymphoblastic leukemia with a hypodiploid karyotype with less than 40 chromosomes: the basis for division into two subgroups. Leukemia 1989; 3:749-752.
15. Mitelman F. Catalog of chromosome aberrations in cancer. 5th ed. New York:Wiley-Liss, 1994.
16. Paietta E, Gucalp R, Wiernik PH. Monosomy 7 in multilineage and acute lymphoblastic leukaemia. Br J Haematol 1991; 79:152-155.
17. Kadam P, Masterson M, Soukup S et al. Detection of unexpected clones of monosomy 7 in childhood acute lymphoblastic leukemia using fluorescence in situ hybridization. Anticancer Research 1994; 14:545-548.
18. Silengo M, Vassallo E, Barisone E et al. Monosomy 20 in childhood acute lymphoblastic leukemia. Cancer Genet Cytogenet 1992; 59:177-179.
19. Betts DR, Kingston JE, Dorey EL et al. Monosomy 20: a nonrandom finding in childhood acute lymphoblastic leukemia. Genes Chromosom Cancer 1990; 2:182-185.

20. Gibbons B, MacCallum P, Watts E et al. Near haploid acute lymphoblastic leukemia: seven new cases and a review of the literature. Leukemia 1991; 5:738-743.
21. Onodera N, McCabe NR, Nachman JB et al. Hyperdiploidy arising from near-haploidy in childhood acute lymphoblastic leukemia. Genes Chromosom Cancer 1992; 4:331-336.
22. Onodera N, McCabe NR, Rubin CM. Formation of a hyperdiploid karyotype in childhood acute lymphoblastic leukemia. Blood 1992; 80:203-208.
23. Pui C-H, Carroll AJ, Raimondi SC et al. Clinical presentation, karyotypic characterization, and treatment outcome of childhood acute lymphoblastic leukemia with a near-haploid or hypodiploid <45 line. Blood 1990; 75:1170-1177.
24. Fletcher JA, Kimball VM, Lynch E et al. Prognostic implications of cytogenetic studies in an intensively treated group of children with acute lymphoblastic leukemia. Blood 1989; 74:2130-2135.
25. Harbott J, Ritterback J, Ludwig WD et al. Clinical significance of cytogenetic studies in childhood acute lymphoblastic leukemia: experience of the BFM trials. Recent Results Cancer Res 1993; 131:123-132.
26. Kobayashi H, Maseki N, Homma C et al. Clinical significance of chromosome abnormalities in childhood acute lymphoblastic leukemia in Japan. Leukemia 1994; 8:1944-1950.
27. GFCH. Groupe Francais de Cytogenetique Hematologique: Collaborative study of karyotypes in childhood acute lymphoblastic leukemias. Leukemia 1993; 7:10-19.
28. Secker-Walker LM, Prentice HG, Durrant J et al. Cytogenetics adds independent prognostic information in adults with acute lymphoblastic leukaemia on MRC trial UKALL XA. Br J Haematol 1997; in press.
29. Pui C-H, Crist WM. Cytogenetic abnormalities in childhood acute lymphoblastic leukemia correlates with clinical features and treatment outcome. Leukemia and Lymphoma 1992; 7:259-275.
30. Secker-Walker LM. Prognostic and biological importance of chromosome findings in acute lymphoblastic leukemia. 10th anniversary article. Cancer Genet Cytogenet 1990; 49:1-13.
31. Raimondi SC, Behm FG, Roberson PK et al. Cytogenetics of childhood T-cell leukemia. Blood 1988; 72:1560-1566.
32. Swansbury GJ, Secker-Walker LM, Lawler SD et al. Chromosomal findings in acute lymphoblastic leukaemia of childhood: An independent prognostic factor. Lancet 1981; ii:249-250.
33. Pui C-H, Carroll AJ, Head D et al. Near-triploid and near-tetraploid acute lymphoblastic leukemia of childhood. Blood 1990; 76:590.

34. Raimondi SC, Roberson PK, Pui CH et al. Hyperdiploid (47-50) acute lymphoblastic leukemia in children. Blood 1992; 79:3245-3252.
35. Moorman AV, Clark R, Farrell DM et al. Probes for hidden hyperdiploidy in acute lymphoblastic leukaemia. Genes Chromosom Cancer 1996; 16:40-45.
36. Hodohara K, Fujiyama Y, Inoue T et al. Trisomy 4 in a case of acute lymphocytic leukemia (L1). [Review]. Cancer Genet Cytogenet 1992; 62:88-91.
37. Moreau P, Talmant P, Milpied N et al. Trisomy 4 associated with acute lymphoblastic leukaemia. Br J Haematol 1991, 78.576.
38. Sandberg AA, Morgan SS, Morgan R et al. Trisomy 5 as sole anomaly in acute lymphoblastic leukemia. Cancer Genet Cytogenet 1988; 36:31-34.
39. Rao PN, Buss D, Brown S et al. Trisomy 5 as the sole abnormality in acute lymphoblastic leukemia. A second case and review. [Review]. Cancer Genet Cytogenet 1994; 75:117-119.
40. Garipidou V, Yamada T, Prentice HG et al. Trisomy 8 in acute lymphoblastic leukemia (ALL): a case report and update of the literature. Leukemia 1990;4:717-719.
41. Pettenati MJ, Rao N, Wofford M et al. Presenting characteristics of trisomy 8 as the primary cytogenetic abnormality associated with childhood acute lymphoblastic leukemia. A Pediatric Oncology Group (POG) Study (8600/8493). Cancer Genet Cytogenet 1994; 75:6-10.
42. Angioni A, Ruscio C, Giovannelli L et al. Characterization of a new case of trisomy 8 in acute lymphoblastic leukemia. Cancer Genet Cytogenet 1994; 74:144-146.
43. Secker-Walker LM, Fitchett M. Constitutional and acquired trisomy 8. Leukemia Research 1995; 19:737-740.
44. Raimondi SC, Pui CH, Head D et al. Trisomy 21 as the sole acquired chromosomal abnormality in children with acute lymphoblastic leukemia. Leukemia 1992; 6(3):171-175.
45. Watson MS, Carroll AJ, Shuster JJ et al. Trisomy 21 in childhood acute lymphoblastic leukemia: a Pediatric Oncology Group study (8602). Blood 1993; 82:3098-3102.
46. Tuszynski A, Dhut S, Young BD et al. Detection and significance of bcr-abl mRNA transcripts and fusion proteins in Philadelphia-positive adult acute lymphoblastic leukemia. Leukemia 1993; 7:1504-1508.
47. Crist W, Carroll A, Shuster J et al. Philadelphia chromosome positive childhood acute lymphoblastic leukemia: clinical and cytogenetic characteristics and treatment outcome. A Pediatric Oncology Group study. Blood 1990; 76:489-494.
48. Ribeiro RC, Abromowitch M, Raimondi SC et al. Clinical and biologic hallmarks of the Philadelphia chromosome in childhood acute lymphoblastic leukemia. Blood 1987; 70:948-953.

49. Walters R, Kantarjian HM, Keating MJ et al. The importance of cytogenetic studies in adult acute lymphocytic leukemia. Am J Med 1990; 89:579-587.
50. Secker-Walker LM, Hawkins JM, Janossy G et al. Independent prognostic significance of karyotype in adults with acute lymphoblastic leukemia and an increase in poor risk chromosome features with advancing age at diagnosis. Blood 1990; 76:319a.
51. Pui CH, Raimondi SC, Dodge RK et al. Prognostic importance of structural chromosomal abnormalities in children with hyperdiploid (greater than 50 chromosomes) acute lymphoblastic leukemia. Blood 1989; 73:1963-1967.
52. Jackson JF, Boyett J, Pullen J et al. Favorable prognosis associated with hyperdiploidy in children with acute lymphocytic leukemia correlates with extra chromosome 6. Cancer 1990; 66:1184-1189.
53. Rubin CM, Le Beau MM, Mick R et al. Impact of chromosomal translocations on prognosis in childhood acute lymphoblastic leukemia. Journal of Clinical Oncology 1991;9:2183-2192.
54. Dastugue N, Robert A, Payen C et al. Prognostic significance of karyotype in a twelve-year follow-up in childhood acute lymphoblastic leukemia. Cancer Genet Cytogenet 1992; 64:49-55.
55. Raimondi SC, Pui C-H, Hancock ML et al. Heterogeneity of hyperdiploid (51-67) childhood acute lymphoblastic leukemia. Leukemia 1996; 10:213-224.
56. Trent JM, Stanbridge EJ, McBride HL et al. Tumorigenicity in human melanoma cell lines controlled by introduction of human chromosome 6. Science 1990; 247:568-571.
57. Harris MB, Shuster JJ, Carroll A et al. Trisomy of leukemic cell chromosomes 4 and 10 identifies children with B-progenitor cell acute lymphoblastic leukemia with a very low risk of treatment failure: a Pediatric Oncology Group study. Blood 1992; 79:3316-3324.
58. Look AT, Roberson PK, Williams DL et al. Prognostic importance of blast cell DNA content in childhood acute lymphoblastic leukemia. Blood 1985; 65:1079-1086.
59. Secker-Walker LM. The prognostic implications of chromosomal findings in acute lymphoblastic leukemia. Cancer Genet Cytogenet 1984; 11:233-248.
60. Whitehead VM, Vuchich MJ, Lauer SJ et al. Accumulation of high levels of methotrexate polyglutamates in lymphoblasts from children with hyperdiploid (greater than 50 chromosomes) B-lineage acute lymphoblastic leukemia: a Pediatric Oncology Group study. Blood 1992; 80:1316-1323.
61. Kaspers GJ, Smets LA, Pieters R et al. Favorable prognosis of hyperdiploid common acute lymphoblastic leukemia may be explained by sensitivity to antimetabolites and other drugs: results of an in vitro study. Blood 1995; 85:751-756.

62. Manabe A, Coustan-Smith E, Behm FG et al. Bone marrow-derived stromal cells prevent apoptotic cell death in B-lineage acute lymphoblastic leukemia. Blood 1992; 79(9):2370-2377.
63. Heerema NA, Argyropoulos G, Weetman R et al. Interphase in situ hybridization reveals minimum residual disease in early remission and return of the diagnostic clone in karyotypically normal relapse of the acute lymphoblastic leukemia. Leukemia 1993; 7:537-543.

CHAPTER 4

Major Translocation Subgroups in Early and Pre-B ALL

The clinical importance of translocations which identify subgroups is due to their impact on prognosis and to their association with other prognostic variables. The genetic consequences of several translocations are known and knowledge of the genetic changes brought about by these chromosomal rearrangements has thrown new light on our understanding of the mechanisms of leukemogenesis. Translocations are now firmly established as crucial events in the etiology of neoplastic growth. Furthermore, molecular techniques enable these translocations to be detected in whole cell populations and in interphase as well as metaphase cells. These techniques provide a sensitive way of detecting chromosomal change when cytogenetic techniques are inappropriate or unavailable. Two of these translocations, t(1;19) and t(4;11), will be considered in this chapter. Each is strongly associated with a particular immunophenotype. Translocation t(1;19) is strongly associated with ALL with pre-B blasts (expressing cytoplasmic immunoglobulin (cIg)). Translocation t(4;11) is found in patients with null-cell ALL with blasts of very early B lineage which are CD19 positive but lack the C-ALL antigen CD10.

t(1;19) AND der (19)t(1;19)(q23;p13)

This translocation was first described in 1984.[1-3] Initially, the breakpoints for this translocation were variously described as 1p21 or q23 and 19p13 or q13. The accepted breakpoints 1q23 and

Heterogeneity of *E2A/PBX1* transcripts in cases with t(1;19)

It is still a matter of debate whether the translocation and its variant form consistently express *E2A/PBX1* transcripts. Two groups have consistently reported the presence of the transcript or rearrangement of *E2A* only in cases with pre-B immunophenotype but not in cases with C-ALL blasts.[20,21] Heterogeneity of the breakpoint is supported by two studies in which not all patients with t(1;19) show *E2A/PBX1* fusion.[20,22] Privitera et al[22] have produced evidence that a gene other than *PBX1* might be involved in some cases. Investigation of the *E2A/PBX1* fusion by RT-PCR and *E2A* rearrangement by Southern blotting identified seven cases with the expected fusion protein and three molecular variants. One of these had 27 additional base pairs inserted in frame at the junction site. The other two cases showed rearrangement of *E2A* but no transcript.[22] Three other groups have demonstrated a more widespread association of the translocation with expression of the transcripts regardless of immunophenotype. Patients with t(1;19) with C-ALL, mature-B cell and even with AML have shown expression of *E2A/PBX1* or rearrangement of the *E2A* gene.[15,16,23-25] We at the Royal Free Hospital have recently seen a patient with mature B ALL with both t(8;14) and t(1;19). The presence of an *E2A/PBX1* transcript was shown using nested RT-PCR which suggested a low level of expression. It was suggested that low level of the transcript may be explained by a modifying effect on its expression of genes from the other fusion products such as, in this case, t(8;14).[16] Coexistence of t(1;19) and t(14;18)(q32;q21) has recently been described in a patient with mature B ALL. In this case, rearrangement of *BCL2* located at 18q21 and, expression of the *E2A/PBX1* transcript, again using nested RT-PCR, were both shown.[15]

Clinical Features

Incidence

This translocation or its derived form is found in between 2% and 9% of children with ALL (see Table 4.1). The discrepancy between the figures of Pui et al[10] and others[9,26] may indicate a geographical or racial effect on incidence.

Table 4.1. Incidence by immunophenotype of t(1;19) and der(19)t(1;19) in childhood ALL

All Cases	Pre B	C-ALL	Refs
10/411(2.4%)	NK	NK	GFCH et al[9]
21/989 (2.1%)	6/70 (9%)	15/643 (2.3%)	Harbott et al[26]
45/737 (8.8%)	41/228 (18%)	4/509 (0.8%)	Pui et al[10]
32/1186 (1.7%)	14/113 (12%)	12/592 (2.0%)	Appendix 2

Age Range, Race and White Blood Cell Count

Clinical features of patients with t(1;19) from recent surveys are shown in Table 4.2. There were significantly more females than males in the study by Secker-Walker et al[7] but this has not been reported elsewhere. The age range of patients reported by Secker-Walker et al was <1-54 years with a median age in children of five years and in adults of 25 years. Studies in the United States have noted a high incidence (up to 41%) of black race among children in this subgroup.[10,27] Reported white blood cell counts range from 0.5 - 505 x 10^9/L. Median WBC from the larger series range from 21-28 x 10^9/L.[5,7,8,10,25,27,28]

Immunophenotype and Incidence

The most notable feature of this translocation has been its strong association with a pre-B immunophenotype with blasts expressing cytoplasmic immunoglobulin (cIg)[1] in a particularly high percentage (>70%) of cells (Secker-Walker and Janossy, unpublished data). Amongst patients with a pre-B immunophenotype, up to 18% have been reported as having this translocation (see Table 4.1).[1-3,6,8,10,26] One study of 22 cases with t(1;19) suggested that apart from cIg the blasts of these cases express a characteristic but uncommon profile of surface antigens, i.e., CD19, CD10, CD9, absence of CD34 and partial absence of CD20.[29] However, it is clear that this translocation is also an important primary abnormality in leukemia with other immunophenotypes. It has been described in up to 2% of C-ALL cases (Table 4.1) and has also been seen in mature B ALL as well as in acute myeloid leukemia.[10,15,26,30,31]

PROGNOSIS

The prognostic implications of this abnormality were initially unclear. An association with poor-risk features was noted by some but not others.[1,2,5,32] Estimated event-free survivals for children with t(1;19) or der(19) taken from recent publications are compared in Table 4.3. One study identified t(1;19) as the major adverse feature in pre-B ALL (see Fig. 4.3),[27] but among pre-B cases those with any translocation, including t(1;19), fared badly compared with those without a translocation.[8] Further studies from St Jude's Children's Hospital showed that more intensive treatment apparently improved the survival of pre-B cases with any translocation (see Fig. 4.4).[8] It may be that prognosis for patients with t(1;19) is poor only among patients with pre-B ALL and not among those with C-ALL.[10] In a large multinational study which considered 66 cases with t(1;19), for which full immunophenotyping was not always available, it was shown that the derived form der(19)t(1;19) was associated with a better prognosis than the balanced form t(1;19) (see Fig. 4.5). The better prognosis for the 36 cases with der(19) (66% survival at five years) than for the 30 cases with t(1;19)(q23;p13) (34% at five years) was statistically significant ($p=0.002$). The impact of the karyotypic variant on prognosis in this survey was in marked contrast to that of age, white blood cell count and additional chromosomal change, none of which were prognostically important.[7] Pui et al[10] have also compared survival of the der or t forms in 41 pre-B and four C-ALL cases. An excess of cases with the der form was even more pronounced. Patients with pre-B ALL showed the der form in 31 cases and only 10 with the balanced translocation. All four cases of C-ALL had the der form. Patients were treated on four different protocols over a 15 year period. When only pre-B cases were considered, the five year event-free survival for cases with the der form was 65% and for t(1;19) cases was 41%. The numbers were few and these differences were not statistically significant. All four C-ALL cases were still in first remission at the time of the report.[10] Adults aged >14

Table 4.2. Clinical features associated with t(1;19) and der(19)t(1;19)

No	Sex M:F	Age/yrs Median (range)	WBC x10^9/L Median range	Race White: non-white	Refs
6	2:4	3.5 (<1-16)	50 (18-164)	NG	Skikano et al[5]
29	NG	Children	25	17:12	Crist et al[27]
66	25:41	(<1-54)	20.9(3.5-480)	59:7	Secker-Walker et al[7]
41	24:17	5.6(1.3-16.4)	28.1(4.3-505)	27:4	Pui et al[10]
17	18:9	9(1-22)	30	NG	Troussard et al[25]

Table 4.3. Event-free survival of children with t(1;19) and/or der(19)

No. Cases	Immunophenotype	Estimated EFS	Refs
29	PreB	45% at 3 yrs	Crist et al[27]
34	PreB+C-ALL	50% at 3 yrs	Crist et al[27]
14	Pre-B	79% at 4 yrs	Raimondi et al[8]
10	Pre-B	53% at 4 yrs	GFCH[9]
14	Pre-B	55% at 4 yrs	Harbott et al[26]

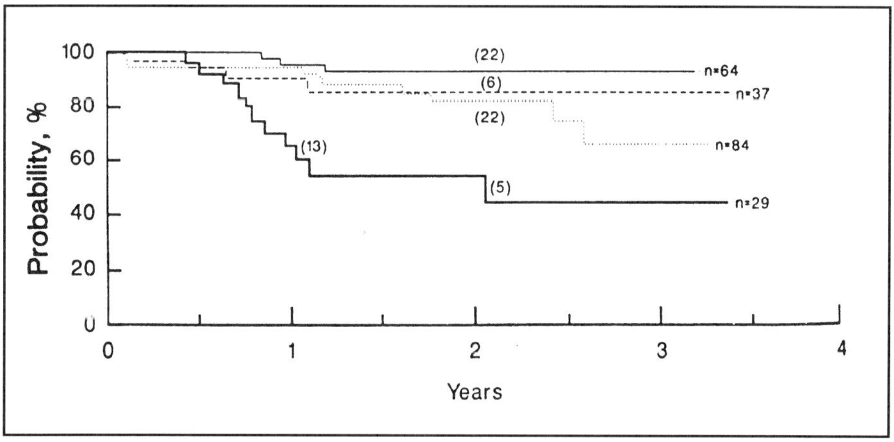

Fig. 4.3. Kaplan Meier estimate of event-free survival for pre-B patients with t(1;19) (bold solid line), with normal karyotype (dotted line), with an identified translocation other than t(1;19) (dashed line) or with another clonal abnormality (faint solid line). Patients with t(1;19) had the worst outcome (p<0.001). Reprinted with permission from Crist et al.[27]

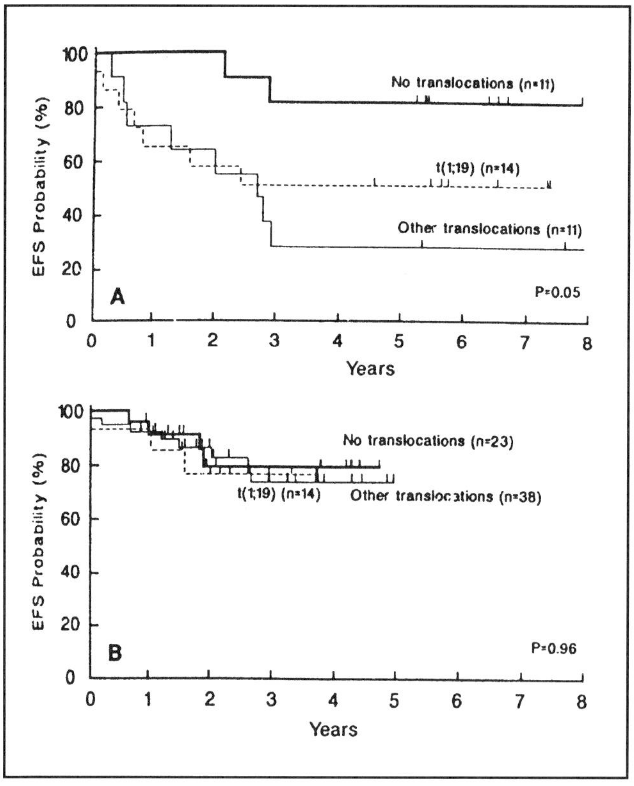

Fig. 4.4. Event-free survival in children with pre-B ALL with t(1;19), other translocations or no translocations in different therapy studies. Survival of t(1;19) patients improved with more recent therapy. Reprinted with permission from Raimondi et al.[8]

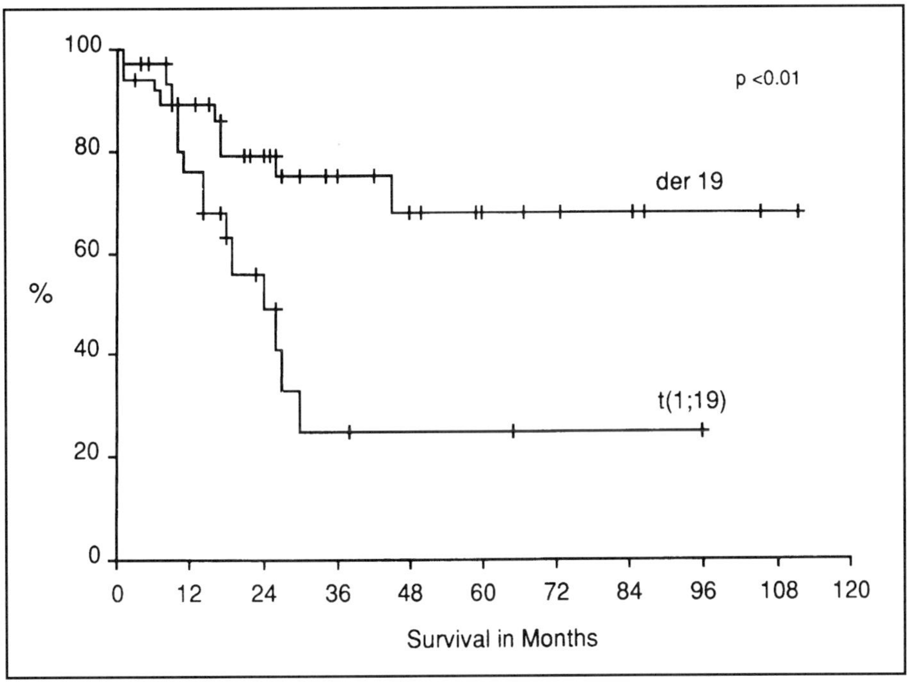

Fig. 4.5. Survival of patients with t(1;19)(q23;p13) and the unbalanced form der(19)t(1;19). Reprinted from Secker-Walker et al.[7]

years on the MRC UKALL XA trail (1985 to 1992) included 10 patients with t or der(19). This subgroup had an event-free survival of 54% at three years. This was considerably better than that for other translocation groups but the number of patients were few and this outcome was not statistically significant.[33]

Molecular Techniques to Identify or Monitor Genetic Change

Rearrangement of the *E2A* gene by Southern blotting and the more sensitive technique of reverse transcription polymerase chain reaction (RT-PCR) can be used to detect or monitor the t(1;19) or the chimeric *E2A/PBX1* transcripts in DNA or RNA respectively from whole cell populations.[17,21,34]

Incidence of t(1;19) ascertained by molecular techniques at diagnosis

Initial results comparing the two techniques for detection of t(1;19) indicate that there may be discrepancies between the cytogenetic and RT-PCR findings. This suggests either that the genetic rearrangement may take place in the absence of visible cytogenetic change, that the t(1;19) clone is not actively dividing or that cytogenetic detection of the clone may fail for technical reasons. In one study, the incidence of the *E2A/PBX1* fusion mRNA as detected by RT-PCR was higher than the incidence of t(1;19) detected cytogenetically. In that study, 21/251 (9.5%) of B cell precursor cases were positive, more than half having escaped detection cytogenetically. On the other hand, 2/12 patients with t(1;19) on cytogenetics did not show the chimeric gene.[35] Devaraj et al have provided evidence that samples from patients with t(1;19) were neither more nor less likely to fail cytogenetics than patients with other chromosomal abnormalities. *E2A/PBX1* fusion was detected by RT-PCR in 3% of patients with failed cytogenetics, giving a similar incidence to that detected overall by cytogenetics. No evidence of *E2A/PBX1* was found among 11 cytogenetically normal samples in which an 11% clone had been excluded with 90% confidence.[36]

Minimal residual disease detection by RT-PCR

The value of RT-PCR for the detection of the clone in remission is at present uncertain. Among 15 patients examined in remission and reported in two publications, eleven became RT-PCR negative in remission. One of these reverted to RT-PCR+ve at 24 months and relapsed four months later. Four patients did not become RT-PCR negative. One of these relapsed five months after testing. Two of the others who were RT-PCR+ at 29 and 36 months have remained well for at least 10 months since.[16,35]

Impact of molecular investigations and FISH

There have been two reports of cases with a supposed t(1;19) which on painting with whole chromosome paints have proved to be otherwise. Both cases had blasts lacking cIg and both were found to have t(1;22) with chromosome 22 mimicking chromosome 19.[37,38] The second case reported had not shown the *E2A/PBX1* fusion transcript and had been reported as supporting evidence for the lack of involvement of these genes in cases of t(1;19) without cIg. It would be of great interest to establish how many other such cases are bona fide t(1;19) cases and whether there is indeed a second set of genes at these breakpoints which occurs in C- rather than pre-B ALL.

t(17;19)(q22;p13)

There is one known variant of t(1;19), namely t(17;19)(q22;p13). This translocation occurs rarely.

MOLECULAR EVENTS

t(17;19) involves fusion of the *E2A* gene on 19 with a gene *HLF* on chromosome 17, a novel basic leucine zipper (bZIP) protein which brings about leukemogenic conversion of a bZIP transcription factor.[39] There are two types of genomic rearrangements resulting in fusion of *E2A* with *HLF*. In type I rearrangements, there is an insertion derived from a cryptic exon at the junction between chromosomes 17 and 19 which includes intronic portions of both *E2A* and *HLF* with intervening nontemplated N nucleotides. Type II rearrangements arise from more 5' breakpoints in *E2A* and result in fusion cDNAs with *E2A* exon 12 spliced directly to the *HLF* exon 4. Both types of rearrangement result in an "in frame" fusion of the two genes.[40,41] It is not clear at present whether heterogeneity of the disease is associated with different breakpoints within the *E2A* and *HLF* genes.

CLINICAL DETAILS

The translocation t(17;19) has been described in at least five patients aged 2-17 (median 15.5) years with leukocyte counts 4.3—71.9 (median 16.1) x 10^9/L and with C-ALL (four cases) or null (one case) immunophenotype. Two of these patients presented with disseminated intravascular coagulation, a rare complication in ALL, that improved with antileukemic therapy. Survival for most patients has been poor.[42,43]

t(4;11)(q21;q23)

This translocation was first seen in 1977 by Oshimura et al[44] in a 27 year old male with null ALL. It was confirmed as recurring change in 1979.[45] An example of the translocation is shown in Figure 4.6. The translocation is accompanied by additional

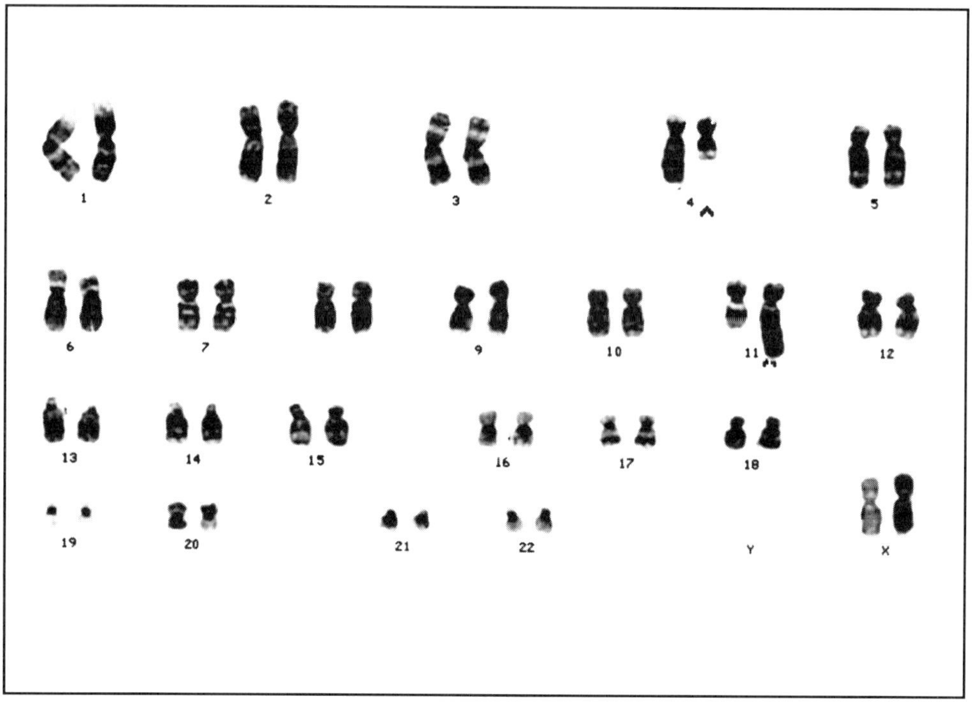

Fig. 4.6. Karyogram of 46,XX,t(4;11)(q21;q23)

change in 30% of cases overall at diagnosis.[46] Additional change is seen least often in infants (18%) and most frequently occurs in 1-9 year olds (54%). In relapse, up to 78% of cases have additional change principally +6, +8, +13 and i(7)(q10q10).[47] In a review article, Pui showed that t(4;11) is usually found in a pseudodiploid clone (90% of cases). Only two of those cases had hyperdiploidy >50, while low hyperdiploidy with 47-50 chromosomes and near tetraploidy accounted for 7.5% and 2.5% of cases, respectively.[48]

Genetic Change

In chromosomal band 11q23 there is a heritable fragile-site. It has been suggested that this fragile-site may be implicated in the etiology of these leukemias.[49] However, there is no evidence for this suggestion. On the contrary, investigation of family members of three infants having ALL and 11q23 breakpoints confirmed the absence of the heritable fragile-site in those families.[50] The gene

Table 4.4. t(4;11) incidence in children and adults

Age group	t(4;11)/Total	Refs
0.3 - 13.5 yrs	6/237 (2.5)	Mirro et al[66]
1 mth - 14 yrs	40/1986 (2.0)	Pui et al[48]
Children	27/989 (2.7)	Harbott et al[26]
Children	2/76 (2.6)	Secker-Walker et al[47]
<16	7/411 (1.7)	GFCH[9]
Children	12/633 (2.0)	Secker-Walker et al[91]
1 - 14 yrs	10/1186 (0.8%)	Appendix 2
1 mth - 80 yrs	3/135 (2.2)	Van der Plas et al[92]
15 - 55 yrs	12/156 (7.7%)	Appendix 2
15 - 74 yrs	10/350(3%)	Secker-Walker et al[33]
15 - 31 yrs	4/48 (8.0)	Levin et al[59]

located at 11q23, which is rearranged in acute leukemias with a breakpoint at 11q23, I shall refer to as *MLL* (myeloid lymphoid leukemia, otherwise known as *ALL-1* or *HRX*). The *MLL* gene spans approximately 100kb of DNA and contains at least 21 exons. It encodes a protein of more than 3910 amino acids containing three regions with homology to sequences within the *Drosophila trithorax* gene. The breakpoint cluster region of *MLL* spans 8 kb. In t(4;11), *MLL* is cleaved within the coding region and results in fusion, in phase, of the open reading frames of *MLL* and the *AF-4* gene on chromosome 4.[51-53] The fusion transcript created on the der(11) chromosome is believed to be the critical gene in the etiology of this type of leukemia since *MLL/AF4* mRNA has been shown in 23/23 childhood cases with t(4;11) ALL. In contrast the reciprocal product *AF4/MLL* from the der(4) was seen in only 85% of cases.[54]

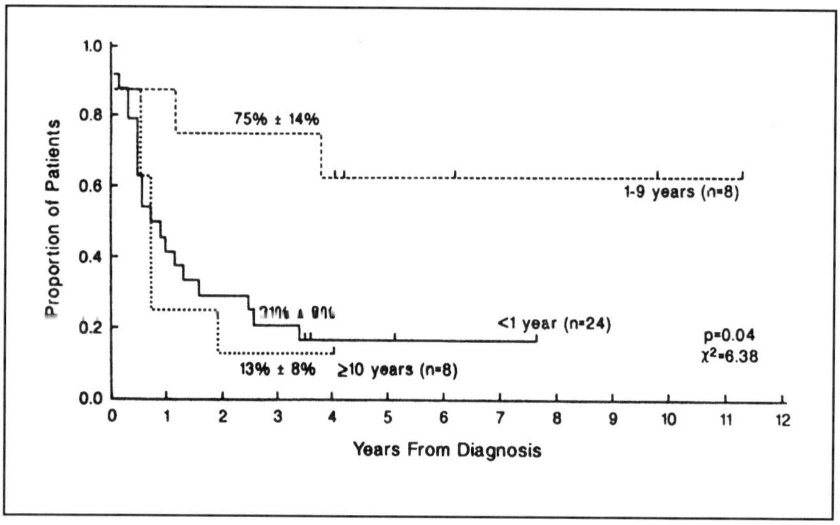

Fig. 4.7. Event-free survival curves for patients with t(4;11) in different age groups. Reprinted from Pui et al. Blood 1994:83, 2384-2385.

change were prognostically important.[48,67] Prognosis for adults with t(4;11) is also extremely poor.[33,57,58]

Incidence of *MLL* Rearrangement and *MLL/AF4* Fusion Products at Diagnosis

Investigation of the *MLL* gene by Southern blotting can be used for the detection of 11q23 abnormalities at diagnosis. Investigation of infant ALL showed rearrangement of *MLL* in 78/96 (81%) cases. 11q23 abnormalities were detected cytogenetically in only 50 cases of infant ALL.[74] Investigation of *MLL* in one adult series showed 2/86 (2.3%) with rearrangement of the gene.[75] The patients were one male aged 43 years and one female aged 64 years, WBC was very high (392 x 10^9/L) in one case only. One of these had been detected as t(4;11) and the other had failed cytogenetically. Survival in each case was two weeks.[75] This incidence is lower than that reported cytogenetically in adult series.[57,58]

The fusion transcript of *MLL/AF4* can be detected by RT-PCR as shown in 15/15 cases with t(4;11), although different breakpoints on chromosome 11 and alternative splicing of *MLL* exon 8 resulted in DNA fragments of different sizes.[76]

VARIANT 11q23 TRANSLOCATIONS

In 1989, Raimondi et al described translocations of 11q23 in childhood ALL to 10 different partner sites. As well as 4q21 translocations were seen to 1q32, 7q11, 9p21, 10p14-15, 12p13, 12q24, 15q12-21 and 19p13. Also seen was an inversion of 11 which resulted in fusion of 11q23 with 11p13.[77] These abnormalities have since been confirmed as recurring abnormalities in ALL and to the list of reciprocal breakpoints have been added 6q27,[78] 16p13, 17q21 and Xq13.[79]

GENETICS OF 11q23

It is now recognized that the *MLL* gene may fuse with a great number of other genes earning it the reputation of "a most promiscuous gene." Hunger et al[80] have shown rearrangement of the *MLL* gene on Southern blotting in 26/27 patients with 11q23 abnormalities to nine different partner sites detected cytogenetically at diagnosis and nine secondary leukemias. Partner genes to *MLL* in ALL have been identified at 1p32 *AF-1p*,[81] 6q27 *AF6*,[78] 9p21-24 *AF9*,[82,83] 10p13-14 *AF10*, 12p13 *TEL*,[84] 14q11 *TCRD*,[85] 17q21 *AF17*,[86] 19p13 *ENL*[87] and Xq13 *AFX*.

CLINICAL FEATURES

Clinical features of patients with variant translocations at 11q23 are similar to those with t(4;11). These patients as a whole were younger, more often black and had higher WBCs than patients without 11q23 involvement.[77]

t(11;19)(q23;p13)

The most widely reported variant to t(4;11) is t(11;19)(q23;p13). This is found more frequently in AML or biphenotypic leukemia but also occurs in approximately 0.5% of childhood ALL.[88] These cases have been reported as clinically similar to those with t(4;11) with high WBCs, CD19+ve, CD10-ve immunophenotype and predominantly in infants and in females.[89,90]

Abnormalities of 11q23 found in infants will be described more fully in chapter 8.

SUMMARY

In summary, two translocations with strong clinical and blast cell associations have been described. The translocation t(1;19)(q23;p13) is chiefly found in blasts of pre-B immunophenotype although it also has been described in several other cell types of both lymphoid and myeloid lineage. Its prognostic implications are not entirely clear. The t(4;11)(q21;q23) is found almost exclusively in pre, pre-B blasts and is associated with poor-risk clinical features. Prognosis is particularly poor in infants and adults. Both translocations have been cloned and probes to detect the translocations by the aberrant fusion proteins they create are available. These have been used to screen diagnostic and remission samples for the presence of t(1;19) and t(4;11) clones.

REFERENCES

1. Carroll AJ, Crist WM, Parmley RT et al. Pre-B cell leukemia associated with chromosome translocation 1;19. Blood 1984; 63:721-724.
2. Michael PM, Levin MD, Garson OM. Translocation 1;19—a new cytogenetic abnormality in acute lymphocytic leukemia. Cancer Genet Cytogenet 1984; 12:333-341.
3. Williams DL, Look AT, Melvin SL et al. New chromosomal translocations correlate with specific immunophenotypes of childhood acute lymphoblastic leukemia. Cell 1984; 36:101-109.
4. Jack I, Seshadri R, Garson M et al. RCH-ACV: a lymphoblastic leukemia cell line with chromosome translocation 1;19 and trisomy 8. Cancer Genet Cytogenet 1986; 19:261-269.
5. Shikano T, Kaneko Y, Takazawa M et al. Balanced and unbalanced 1;19 translocation-associated acute lymphoblastic leukemias. Cancer 1986; 58:2239-2243.
6. Garry MB, Grindley RM. Translocation between chromosomes 1 and 19 in acute lymphoblastic leukemia. Cancer Genet Cytogenet 1984; 13:343-344.
7. Secker-Walker LM, Berger R, Fenaux P et al. Prognostic significance of the balanced t(1;19) and unbalanced der(19)t(1;19) translocations in acute lymphoblastic leukemia. Leukemia 1992; 6:363-369.
8. Raimondi SC, Behm FG, Roberson PK et al. Cytogenetics of pre-B-cell acute lymphoblastic leukemia with emphasis on prognostic implications of the t(1;19). J Clin Oncol 1990; 8:1380-1388.
9. GFCH. Groupe Francais de Cytogenetique Hematologique: Collaborative study of karyotypes in childhood acute lymphoblastic leukemias. Leukemia 1993; 7:10-19.

10. Pui CH, Raimondi SC, Hancock ML et al. Immunologic, cytogenetic, and clinical characterization of childhood acute lymphoblastic leukemia with the t(1;19) (q23; p13) or its derivative. Journal of Clinical Oncology 1994; 12:2601-2606.
11. Shippey CA, Layton M, Secker-Walker LM. Leukemia characterized by multiple sub-clones with unbalanced translocations involving different telomeric segments: Case report and review of the literature. Genes Chromosom Cancer 1990; 2:14-17.
12. Griffin TC, Tomlinson GE, Raimondi SC et al. Childhood acute lymphoblastic leukemia with both t(1;19) and t(9;22). Leukemia 1992; 6:535-540.
13. Kouides PA, Phatak PD, Wang N et al. B-cell acute lymphoblastic leukemia with L1 morphology and coexistence of t(1;19) and t(14;18) chromosome translocations. Cancer Genet Cytogenet 1994; 78:23-27.
14. Wlodarska I, Stul M, De Wolf-Peeters C et al. t(1;19) without detectable E2A rearrangements in two t(14;18)-positive lymphoma/leukemia cases. Genes Chromosom Cancer 1994; 10:171-176.
15. Rowe D, Devaraj PE, Irving JAE et al. A case of mature B-cell ALL with coexistence of t(1;19) and t(14;18) and expression of the E2A/PBX1 fusion gene. Br J Haematol 1996; 94:133-135.
16. Devaraj PE, Foroni L, Janossy G et al. Expression of the E2A-PBX1 fusion transcripts in t(1;19)(q23;p13) and der(19)(1;19) at diagnosis and in remission of acute lymphoblastic leukemia with different B lineage immunophenotypes. Leukemia 1995; 9:821-825.
17. Mellentin JD, Murre C, Donlon TA et al. The gene for enhancer binding proteins E12/E47 lies at the t(1; 19) breakpoint in acute leukemias. Science 1989; 246:379-382.
18. Nourse J, Mellentin JD, Galili N et al. Chromosomal translocation t(1;19) results in synthesis of a homeobox fusion mRNA that codes for a potential chimeric transcription factor. Cell 1990; 60:535-545.
19. Uckun FM, Downing JR, Gunther R et al. Human t(1;19)(q23;p13) pre-B acute lymphoblastic leukemia in mice with severe combined immunodeficiency. Blood 1993; 81:3052-3062.
20. Privitera E, Kamps MP, Hayashi Y et al. Different molecular consequences of the 1;19 chromosomal translocation in childhood B-cell precursor acute lymphoblastic leukemia. Blood 1992; 79:1781-1788.
21. Hunger SP, Galili N, Carroll AJ et al. The t(1;19)(q23;p13) results in consistent fusion of E2A and PBX1 coding sequences in acute lymphoblastic leukemias. Blood 1991; 77:687-693.
22. Privitera E, Luciano A, Ronchetti D et al. Molecular variants of the 1;19 chromosomal translocation in pediatric acute lymphoblastic leukemia (ALL). Leukemia 1994; 8:554-559.

23. Izraeli S, Henn T, Strobl H et al. Expression of identical E2A/PBX1 fusion transcripts occurs in both pre-B and early pre-B immunological subtypes of childhood acute lymphoblastic leukemia. Leukemia 1993; 7:2054-2056.
24. Yamada T, Craig JM, Hawkins JM et al. Molecular investigation of 19p13 in standard and variant translocations: the E12 probe recognizes the 19p13 breakpoint in cases with t(1;19) and acute leukemia other than pre-B immunophenotype. Leukemia 1991; 5:36-40.
25. Troussard X, Rimokh R, Valensi F et al. Heterogeneity of t(1;19)(q23;p13) acute leukaemias. Br J Haematol 1995; 89:516-526.
26. Harbott J, Ritterback J, Ludwig WD et al. Clinical significance of cytogenetic studies in childhood acute lymphoblastic leukemia: experience of the BFM trials. Recent Results Cancer Res 1993;131:123-132.
27. Crist WM, Carroll AJ, Shuster JJ et al. Poor prognosis of children with pre-B acute lymphoblastic leukemia is associated with the t(1;19)(q23;p13): A Pediatric Oncology Group study. Blood 1990; 76:117-122.
28. Raimondi S, Roberson P, Behm FG et al. Clinical significance of the t(1;19)(q23;p13) in childhood acute lymphoblastic leukemia. Proc Am Ass Cancer Res 1990; 31:26.
29. Borowitz MJ, Hunger SP, Carroll AJ et al. Predictability of the t(1;19)(q23;p13) from surface antigen phenotype: implications for screening cases of childhood acute lymphoblastic leukemia for molecular analysis: a pediatric oncology group study. Blood 1993; 82:1086-1091.
30. Lessard M, Fenneteau O, Sainty D et al. Translocation t(1;19) in acute lymphoblastic leukemia patients with cytological presentation simulating L3-ALL (Burkitt-like). Leukemia & Lymphoma 1993; 11:149-152.
31. Brito Babapulle F, Secker-Walker LM, Shippey C et al. Myelodysplasia and acute myeloid leukaemia following bone marrow transplantation for acute lymphoblastic leukaemia: secondary leukaemia or stem-cell relapse? Br J Haematol 1989; 73:417-419.
32. Lai JL, Fenaux P, Estienne MH et al. Translocation t(1;19)(q23;p13) in acute lymphoblastic leukemia. A report on six new cases and an unusual t(17;19)(q11;q13), with special reference to prognostic factors. Cancer Genet Cytogenet 1989; 37:9-17.
33. Secker-Walker LM, Prentice HG, Durrant J et al. Cytogenetics adds independent prognostic information in adults with acute lymphoblastic leukaemia on MRC trial UKALL XA. Br J Haematol 1997; in press.

34. Kamps MP, Murre C, Sun XH et al. A new homeobox gene contributes the DNA binding domain of the t(1;19) translocation protein in pre-B ALL. Cell 1990; 60:547-555.
35. Izraeli S, Janssen JWG, Haas OA et al. Detection and clinical relevance of genetic abnormalities in pediatric acute lymphoblastic leukemia: a comparison between cytogenetic and polymerase chain reaction analyses. Leukemia 1993; 7:671-678.
36. Devaraj PE, Foroni L, Kitra-Roussos V et al. Detection of BCR-ABL and E2A-PBX1 fusion genes by RT-PCR in acute lymphoblastic leukaemia with failed or normal cytogenetics. Br J Haematol 1995; 89:349-355.
37. Rieder H, Kolbus U, Koop U et al. Translocation t(1;22) mimicking t(1;19) in a child with acute lymphoblastic leukemia as revealed by chromosome painting. Leukemia 1993; 7:1663-1666.
38. Filatov LV, Behm FG, Pui CH et al. Childhood acute lymphoblastic leukemia with equivocal chromosome markers of the t(1;19) translocation. Genes Chromosom Cancer 1995; 13:99-103.
39. Hunger SP, Ohyashiki K, Toyama K et al. Hlf, a novel hepatic bZIP protein, shows altered DNA-binding properties following fusion to E2A in t(17;19) acute lymphoblastic leukemia. Genes & Development 1992; 6:1608-1620.
40. Hunger SP, Devaraj PE, Foroni L et al. Two types of genomic rearrangements create alternative E2A-HLF fusion proteins in t(17;19)-ALL. Blood 1994; 83:2970-2977.
41. Inaba T, Roberts WM, Shapiro LH et al. Fusion of the leucine zipper gene HLF to the E2A gene in human acute B-lineage leukemia. Science 1992; 257:531-534.
42. Devaraj PE, Foroni L, Sekhar M et al. E2A/HLF fusion cDNAs and the use of RT-PCR for the detection of minimal residual disease in t(17;19)(q22;p13) acute lymphoblastic leukemia. Leukemia 1994; 8:1131-1138.
43. Raimondi SC, Privitera E, Williams DL et al. New recurring chromosomal translocations in childhood acute lymphoblastic leukemia. Blood 1991; 77:2016-2022.
44. Oshimura M, Freeman AI, Sandberg AA. Chromosomes and causation of human cancer and leukemia. XXVI. Banding studies in acute lymphoblastic leukemia (ALL). Cancer 1977;40:1161-1172.
45. Van den Berghe H, Parloir C, David G et al. A new characteristic karyotypic anomaly in lymphoproliferative disorders. Cancer 1979; 44:188-195.
46. Pui C-H. Acute leukemias with the t(4;11)(q21;q23). Leukemia and Lymphoma 1992; 7:173-179.

74. Rubnitz JE, Link MP, Shuster JJ et al. Frequency and prognostic significance of HRX rearrangements in infant acute lymphoblastic leukemia: a Pediatric Oncology Group study. Blood 1994; 84:570-573.
75. Stock W, Thirman MJ, Dodge RK et al. Detection of MLL gene rearrangements in adult acute lymphoblastic leukemia. A Cancer and Leukemia Group B study. Leukemia 1994; 8:1918-1922.
76. Biondi A, Rambaldi A, Rossi V et al. Detection of ALL-1/AF4 fusion transcript by reverse transcription-polymerase chain reaction for diagnosis and monitoring of acute leukemias with the t(4;11) translocation. Blood 1993; 82:2943-2947.
77. Raimondi SC, Peiper SC, Kitchingman GR et al. Childhood acute lymphoblastic leukemia with chromosomal breakpoints at 11q23. Blood 1989; 73:1627-1634.
78. Kobayashi H, Espinosa R, Thirman MJ et al. Heterogeneity of breakpoints of 11q23 rearrangements in hematologic malignancies identified with fluorescence in situ hybridization. Blood 1993; 82:547-551.
79. Mitelman F. Catalog of chromosome aberrations in cancer. 5th ed. New York: Wiley-Liss, 1994.
80. Hunger SP, Tkachuk DC, Amylon MD et al. HRX involvement in de novo and secondary leukemias with diverse chromosome 11q23 abnormalities [see comments]. Blood 1993; 81:3197-3203.
81. Bernard OA, Mauchauffe M, Mecucci C et al. A novel gene, AF-1p, fused to HRX in t(1;11)(p32;q23), is not related to AF-4, AF-9 nor ENL. Oncogene 1994; 9:1039-1045.
82. Fan YS, Rowe JM, Dal Cin PD et al. A translocation t(9;11)(p11;q23) in T-cell acute lymphoblastic leukemia (FAB-12). Cancer Genet Cytogenet 1988; 31:263-269.
83. Lida S, Seto M, Yamamoto K et al. MLL T3 gene on 9p22 involved in t(9;11) leukemia encodes a serine/proline rich protein homologous to MLLT1 on 19p13. Oncogene 1993; 8:3085-3092.
84. Jani Sait SN, Raimondi SC, Look AT et al. A t(11;12) 11q23 leukemic breakpoint that disrupts the MLL gene. Genes Chromosom Cancer 1993; 7:28-31.
85. Burnett RC, Espinosa R, Shows TB et al. Molecular analysis of a t(11;14)(q23;q11) from a patient with null-cell acute lymphoblastic leukemia. Genes Chromosom Cancer 1993; 7:38-46.
86. Prasad R, Leshkowitz D, Gu Y et al. Leucine-zipper dimerization motif encoded by the AF17 gene fused to ALL-1(MLL) in acute leukemia. Proc Natl Acad Sci USA 1994; 91:8107-8111.
87. Rubnitz JE, Morrissey J, Savage PA et al. ENL, the gene fused with HRX in t(11;19) leukemias, encodes a nuclear protein with

transcriptional activation potential in lymphoid and myeloid cells. Blood 1994; 84:1747-1752.
88. Hudson MM, Raimondi SC, Behm FG et al. Childhood acute leukemia with t(11;19) (q23;p13). Leukemia 1991; 5:1064-1068.
89. Huret JL, Brizard A, Slater R et al. Cytogenetic heterogeneity in t(11;19) acute leukemia: clinical, hematological and cytogenetic analyses of 48 patients—updated published cases and 16 new observations. [Review]. Leukemia 1993; 7:152-160.
90. Morgan GJ, Cotter F, Katz FE et al. Breakpoints at 11q23 in infant leukemias with the t(11;19)(q23; p13) are clustered. Blood 1992; 80:2172-2175.
91. Secker-Walker LM. Prognostic and biological importance of chromosome findings in acute lymphoblastic leukemia. 10th anniversary article. Cancer Genet Cytogenet 1990; 49:1-13.
92. Van der Plas DC, Hahlen K, Hagemeijer A. Prognostic significance of karyotype at diagnosis in childhood acute lymphoblastic leukemia. Leukemia 1992; 6:176-184.

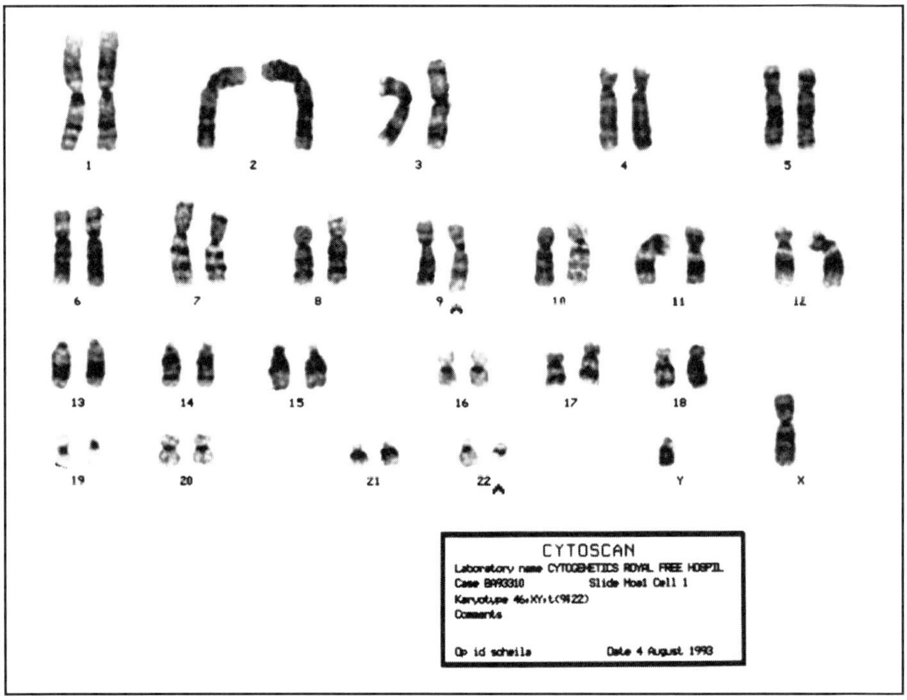

Fig. 5.1. Karyogram with 46,XY,t(9;22)(q34;q11).

immediately following these concluded that such cases in children and adults were in reality CGL in lymphoid blast crisis masquerading as ALL.[8-11] But in 1977, Bloomfield et al established Ph positive ALL (Ph+ALL) as an important independent subgroup of ALL.[12] Ph+ALL and CGL in lymphoid blast crisis were recognized as distinct diseases with different clinical presentations. However, in terms of the Ph translocation, morphology of the blasts and clinical features they are indistinguishable from one another. There are several ways in which Ph+ALL may be distinguished from CGL in blast crisis, but each of these apply in only a minority of cases. Approximately 10% of Ph+ALL patients have a bimodal population of Ph+ and chromosomally normal cells at diagnosis while >99% patients with CGL in blast crisis will have 100% Ph+ cells. These normal metaphases are almost certainly part of the population of residual normal marrow elements. Ph+ALL patients may achieve complete hematological remission unlike patients with CGL in blast crisis who, at best, revert to a chronic phase.[13] Patients with Ph+ALL achieve cytogenetic remission much more readily than do patients with CGL. From a total of 32 patients tested in remission, 19 showed total elimination and seven showed partial elimination of the Ph clone.[14,15]

Additional Chromosomal Change

Chromosomal change additional to the Ph translocation at diagnosis has been identified in up to 60% of children and between 50% and 90% of adults with ALL.[16-18] In UKALLXA, additional change was seen at diagnosis in 65% of Ph+ adults.[19] The incidence of change additional to the Ph in ALL is probably somewhat lower than that in CGL at blast crisis. Particularly important are the Ph+ high hyperdiploid clones which account for up to 11% of childhood Ph cases[17,20] and for up to 36% of adults with Ph+ALL.[14,19] The advent of additional change to the derived chromosomes 9 and 22 involved in the translocation as deletion or translocation of the short arm of der(9) and more rarely, additional change to the der(22) as i(22)(q10q10) or add(22)(p) has been described.[21,22]

Ph+ALL with monosomy 7 appears to represent a distinct subgroup of Ph+ALL with a particularly adverse outcome.[23] The Pediatric Oncology group in the U.S. (POG) found 58 Ph+ve children out of 2519 karyotyped over an 8 year period (2.3%) of which 13 (23%) also had partial or total monosomy 7.[17] Ph+ clones are most frequently pseudodiploid. But chromosomal loss or gain also occur and result in hypodiploid or hyperdiploid clones (see Table 5.1). Most additional

Table 5.1. Ploidy and Ph in ALL

	Total cases	HO	PS	47-50	>50	Refs
Children	58	15	34	5	4	Crist et al[17]
Children	18	0	12	2	4	Ribeiro et al[20]
15-75	40	1	29	4	6	Secker-Walker et al[19]
15-75	14	1	8*		5	Tuszynski et al[14]

* 3 cases were bimodal with 47-50 (1 case) and >50 (2 cases)

change appears to have no prognostic or clinical impact (personal observation).

Genetic Change

The Ph translocation juxtaposes the oncogene *ABL* normally located on chromosome 9 at 9q34 with the *BCR* gene on chromosome 22 at 22q11 to produce a chimeric *BCR/ABL* gene. The breakpoint on 9q34 is between the first and second exons (a1 and a2) of the *ABL* oncogene but the site of the breakpoint in the *BCR* gene at 22q11 differs. In 50-70% of Ph+ acute leukemias, the *BCR* gene breaks between exons 1 and 2 (e1 and e2), also known as the minor breakpoint cluster region or *m-BCR* to form an *e2/a2* chimeric gene with a p190 mRNA. In the remaining acute leukemias and in 99% of cases of CGL breakage is between exons 2 and 3 or exons 3 and 4 (b2 b3, b4) of the major BCR or *M-BCR* forming b2/a2 or b3/a2 junctions. Both junctions give rise to a p210 mRNA. These mRNAs are both larger than the normal *ABL* product (p145 mRNA) and have markedly enhanced tyrosine kinase activity[24] (Fig 5.2).

Incidence and Age

Ph+ALL has been found in between 1%-5% of childhood ALL and in between 11% and 30% of adults. The incidence from recent publications is shown in Table 5.2. The figures from two U.K. sources are considerably lower than those from the U.S.[8,10,12,15,17,20,22,25-28] It is important, however, to realize that Ph+ALL increases with increasing age. First of all, two childhood series have shown that the median age of Ph+ patients (7.9 yrs and 9.6 yrs) was significantly older than that of remaining patients (4.8 years) in both series ($p = 0.05$ and $p < 0.001$, respectively).[17,20] Santana et al found a lesser incidence (14/383 or 3.7%) among 1-9 year olds than among 10-20 year olds (9/127 or 7.0%).[29] Bloomfield et al remarked on the greater age of Ph+ adults compared with that of others with ALL.[12] In a study at the Royal Free Hospital, we have shown that the mean age of 23 Ph+ adults was significantly older (39.7 years) than that of the remaining 90 patients with ALL (28.3 years) karyotyped by us ($p < 0.001$). In our study, the incidence of Ph+ALL was shown to increase from

Color Figures

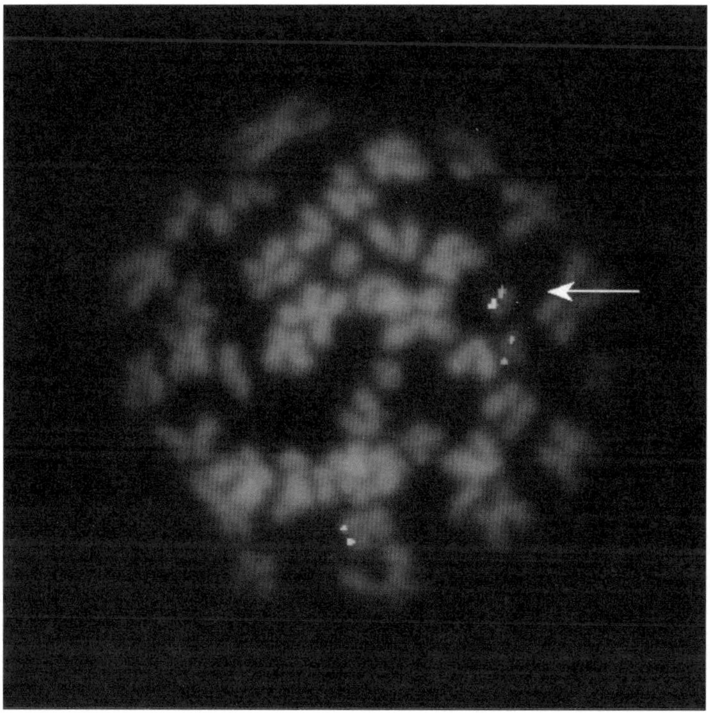

Fig. 5.3. FISH to show the juxtaposition of ABL *and* BCR *genes on chromosome 22, which appears normal. Reproduced from van Rhee et al.*[40]

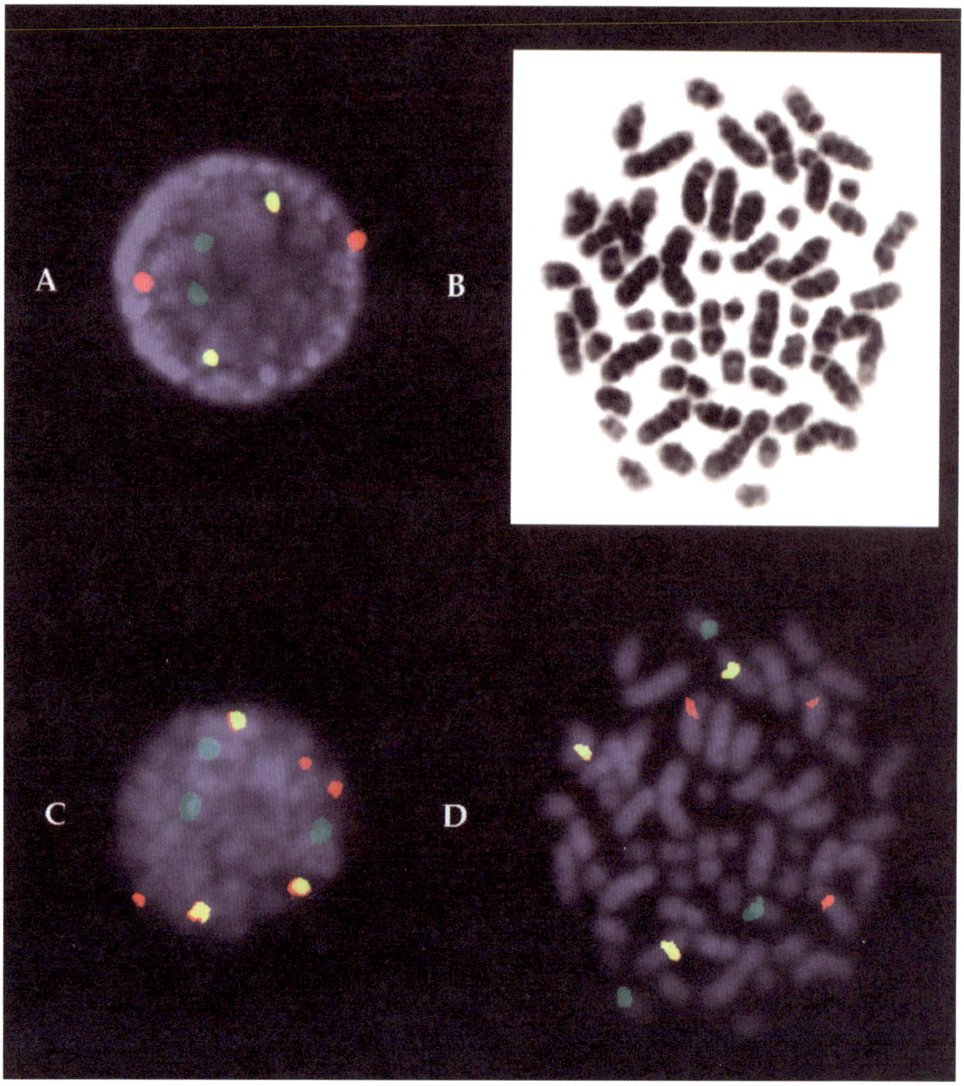

Fig. 9.2 Three color FISH using alpha satellite probes to the centromeric region of chromosomes X (yellow), 6 (red) and 18 (green). A: normal female cell 46, XX, with two copies of each cetromere B: metaphase spread of a hyperdiploid cell with gains of chromosomes X, 6 and 18 showing the G-banding pattern, C: interphase hyperdiploid cell from the same patient showing three copies each of chromosomes X, 6 and 18, D: metaphase spread shown in B showing the probes located at the centromeres. (Kindly donated by Arek Kasprzyk.)

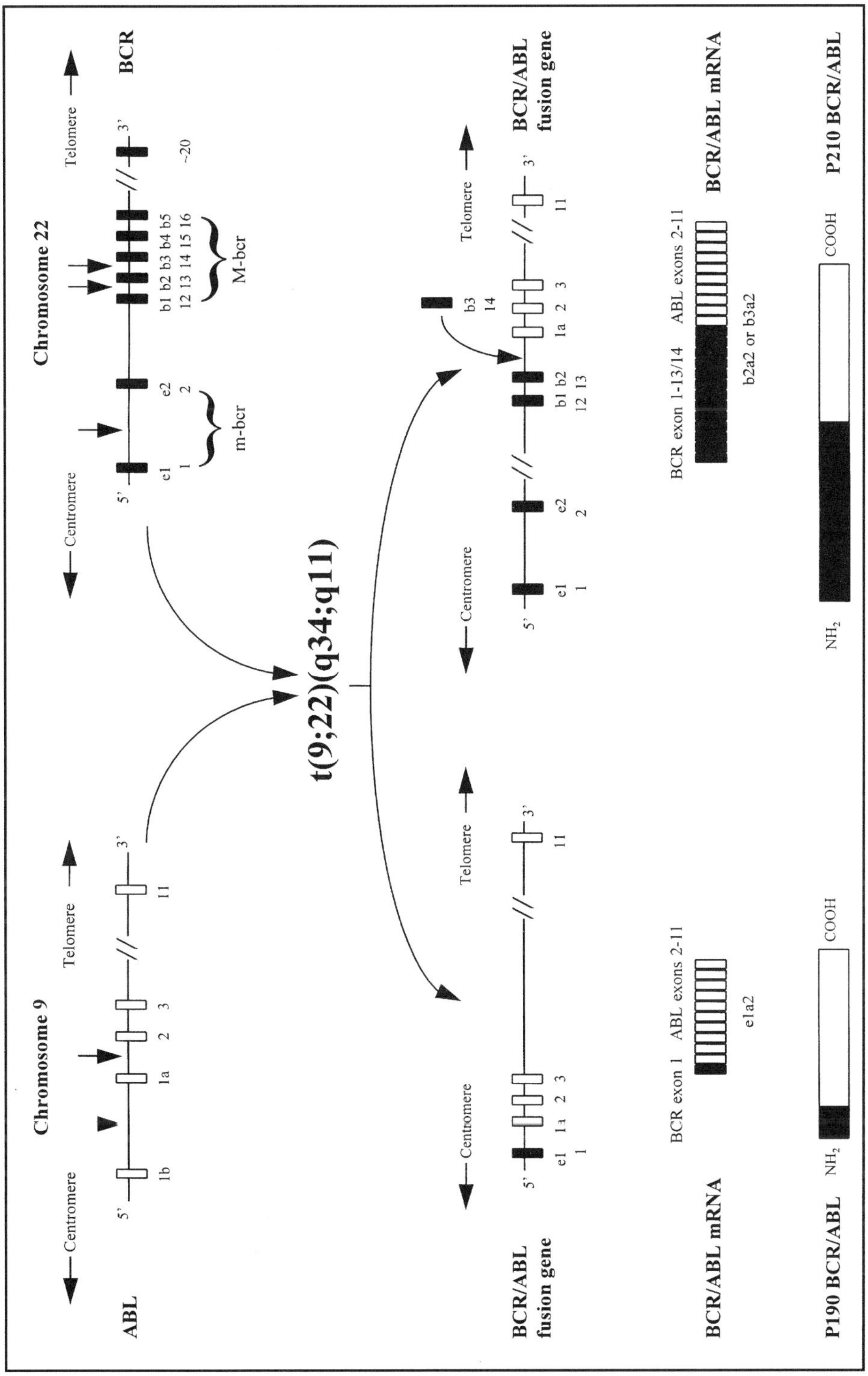

Fig. 5.2. Diagram of genes ABL and BCR and the genetic consequences of t(9;22), according to the breakpoint location in the BCR gene.

Table 5.2. The incidence of t(9;22) in childhood and adult ALL

Age group	Incidence	Refs
<16 yrs	9/411 (2.2%)	GFCH[30]
Children	17/633 (2.7%)	Secker-Walker et al[54]
2-18 yrs	18/366 (4.9%)	Ribeiro et al[20]
Children	58/2519 (2.3%)	Crist et al[17]
19-51	24/108 (22%)	Annino et al[15]
15-75	40/360 (11%)	Secker-Walker et al[19]
15-55	21/156 (13%)	Appendix 2
15-75 yrs	14/70 (20%)	Tuszynski et al[14]

10% at ages 15-20 years to 19% between 21-30 years to 24% between 31-50 years and to 44% over the age of 50 years.[22] Similarly in UKALLXA, the incidence of Ph+ patients increased with age.[19] It follows from these observations that the different incidence of Ph+ALL between series may be explained by the ages of patients in the survey. For example, the 5% Ph+ childhood cases seen by Ribeiro et al[20] included patients up to the age of 18 years while the 2.2% incidence in the GFCH study included children only up to the age of 15 years.[30]

Clinical Features

Patients present with WBCs which range from 3 to >100 x 10^9/L. Higher WBCs are found in childhood Ph+ALL than in other childhood patients (33 x 10^9/L vs 12 x 10^9/L, p = 0.002).[17] This marked difference is not necessarily seen in adults.[22] Blast cell immunophenotypes reported for 68 children in two series shows that 74% of patients had early B-lineage (C-ALL), 16% had a pre-B immunophenotype and 10% had T ALL.[17,20] Ribeiro et al[20] reported that Ph+ patients were more likely to have CNS disease at presentation but this was not seen in a study by Crist et al.[17]

Table 5.3. Remission achievement and prognosis of patients with t(9;22)

Age Group	No. Cases	CR achieved	EFS	Median CR	Refs
<16 yrs	9		14.3% at 4 yrs	10m	GFCH[30]
Children	24		15% at 5 yrs	30m	Harbott et al[32]
2-18 yrs	18	72%	20% at 3 yrs	12m	Ribeiro et al[20]
Children	58	78%			Crist et al[17]
Adults	25	76%	6% at 3.3 yrs	13m	Gotz et al[18]
19-51	24	96%	39% at 3 yrs	20m	Annino et al[15]
15-75 yrs	40	83%	13% at 3 yrs	12m	Secker-Walker et al[19]

CR = complete remission
EFS = event-fee survival

Remission Achievement and Prognosis

Ph+ALL has a poor prognosis. Table 5.3 compares the percentage of patients achieving complete remission times of event-free survival and median survival from recent surveys. Children with Ph+ALL show remission achievement of between 72% and 78%, compared with 90-99% of children in other chromosomal subgroups. In approximately 60% of patients, remission is accompanied by loss of the Ph+ clone.[14,15,17,31] Reported median event-free survivals of Ph+ childhood patients range from 10-30 months and event-free survival by five years is no more than 15%.[20,30,32] The prognosis for adults with Ph+ALL is even worse. Remission achieved ranges from 76% to 96%, but median event-free survival at three years is generally <14%.[18,19] One recent publication showed 39% Ph+ adults event-free at three years.[15] Occasional cases of Ph+ALL with survival >5 years have been reported.[31]

Incidence of *BCR/ABL* Fusion in ALL

Studies in which Ph+ALL patients have been identified using the reverse transcriptase polymerase chain reaction (RT-PCR) for

Table 5.4. *BCR/ABL* with p190 or p210 mRNA in childhood ALL

Cases tested	*BCR/ABL* +ve	p190	p210	Refs
251	3	2	1	Izraeli et al[34]
141	6	5	1	Borkhardt et al[35]
87	5	-	-	Maurer et al[33]
479	14 (2.9%)	7/9 (78%)	2/9 (22%)	Total

detection of p190 or p210 mRNA have reported widely differing figures both for the incidence of *BCR/ABL* ALL as a whole and for the relative frequency of p210 and p190 cases (Tables 5.4, 5.5 and 5.6). In childhood ALL, the incidence of *BCR/ABL* positive patients detected by RT-PCR ranges from 1.2% to 5.7%. The mean figure from three separate studies, however, was 2.9% which is similar to that detected cytogenetically (see Table 5.4).[33-35] The breakpoint location (*m-BCR* or *M-BCR*) has been reported in very few Ph+ children but it appears that *m-BCR* occurs more frequently (Table 5.4). In adults, the incidence of *BCR/ABL* positive patients with ALL has been consistently higher, ranging from 23% to 43% with a mean figure from five surveys of 34% (see Table 5.5). The figure in at least one of these surveys was recognized by the authors as being unnaturally high. Since this was a retrospective study restricted to patients with a high WBC at diagnosis from which cells had been stored.[33] Adults also show an excess of p190 to p210 (66% to 34%) but this is less marked than that in childhood (see Table 5.5). Reports of Ph+ patients investigated for the *M* or *m-BCR* breakpoints in children and adults together are shown in Table 5.6. A total of 93 cases showed 55% with *m-BCR*, 41% with *M-BCR* and 4% with both.[15,26-28,36,37]

Application of Molecular Techniques to Variant Ph Clones and to Cytogenetically Failed or Normal Populations

Molecular analysis is an important addition to the investigation of ALL cases with what may be a variant Ph translocation. At least two such cases with t(2;22) have been shown not to involve rearrangement of either the *M-BCR* or *m-BCR* raising the possibility that some so-called variant Philadelphia translocations may not, in fact, be so.[38,39]

Although it is likely to be rare, chromosomally normal *BCR/ABL* + patients clearly occur in ALL as in CGL. We have recently reported the presence of *M-BCR/ABL* fusion mRNA identified by RT-PCR in a patient with 50 normal metaphases. FISH using probes to *BCR* and *ABL* showed that these genes were juxtaposed on the normal chromosome 22 in most cells thereby confirming the presence of hidden Ph rearrangement (Fig. 5.3).[40]

RT-PCR analysis of one group of 58 adults with Ph-negative ALL, of which 47 had normal cytogenetics, revealed no evidence of *BCR/ABL* fusion genes.[41] However, in cases where cytogenetics fails, molecular analyses may be informative. In a study at the Royal Free Hospital, we have recently shown that the incidence of Ph+ve patients among cases with failed cytogenetics was similar to that

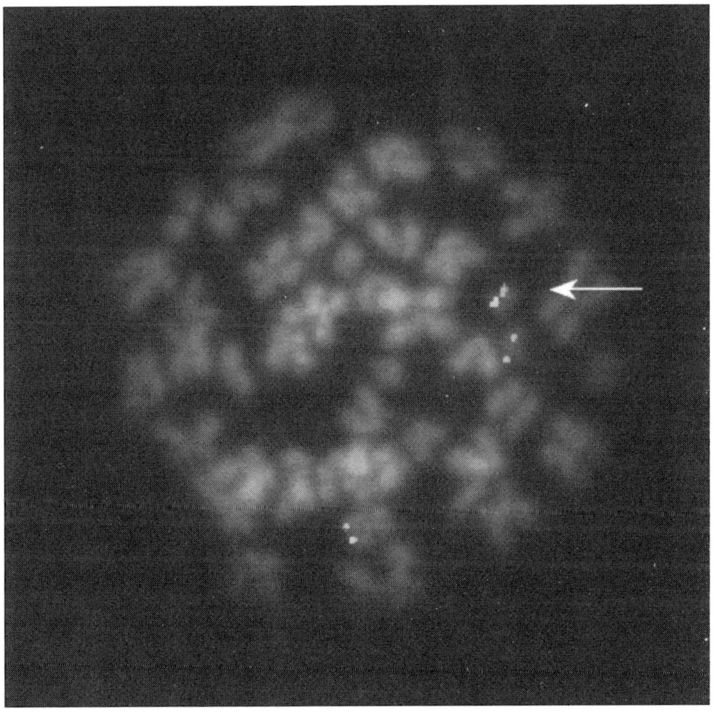

Fig. 5.3. FISH to show the juxtaposition of ABL *and* BCR *genes on chromosome 22, which appears normal. Reproduced from van Rhee et al.*[40] *See color figure in insert.*

Table 5.5. **BCR/ABL** ***with p190 or p210 mRNA in adult ALL***

Cases tested	*BCR/ABL* +ve	p190	p210	Refs
179	77	52	25	Maurer et al[33]
93	21	7	14	Radich et al[26]
56	17	13	4	Westbrook et al[27]
39	13	11	2	Preudhomme et al[28]
44	10	8	2	Tuszynski et al[14]
411	138 (34%)	91 (66%)	47 (34%)	Total

detected in cytogenetic series. Bone marrow samples from 43 patients with ALL in whom cytogenetics had either failed (32 cases) or shown only normal metaphases on 20 cells analyzed (11 cases) revealed six with *BCR/ABL* fusion. These were found only among the failed cases. Four patients had *m-BCR/ABL*, one had *M-BCR/ABL* and one showed both transcripts. These patients were older, had higher WBCs and a significantly worse prognosis than patients without *BCR/ABL* transcripts.[42]

RT-PCR to Monitor Ph+ve Clones for Minimal Residual Disease

Following treatment and achievement of remission, the level of the Ph positive clone frequently, but not always, drops below the level of cytogenetic detection (<5%). The demonstration of p190 or p210 fusion genes at diagnosis provides a valuable and more sensitive means than cytogenetics of monitoring minimal residual disease in remission. The dismal prognosis for Ph+ve ALL patients suggests that the clone is seldom completely eradicated by chemotherapy alone. Monitoring patients who have achieved cytogenetic remission has given heterogeneous results. In some cases, RT-PCR negativity may be followed by long term survival, but the remainder show persistence of the aberrant mRNA or apparent elimination followed by return of the clone and relapse.[36,37] The possibility that quantitative PCR may be a more accurate guide to disease status has to be explored.[43]

Table 5.6. Incidence of* M-BCR *and* m-BCR *breakpoints in adults and children with Ph+ve ALL

BCR/ABL + cases	p190	p210	both	Refs
14	5	5	4	Gehly et al[36]
15	12	3		Myamura et al[37]
17	13	4		Westbrook et al[27]
13	11	2		Preudhomme et al[28]
13	3	10		Annino et al[15]
21	7	14		Radich et al[26]
93	51 (55%)	38 (41%)	(4%)	Total

BCR Breakpoint and Clinical Impact

Clinically, the site of the breakpoint, M or m, apart from the association of *M-BCR* with a higher leukocyte count, appears to have little relevance.[14,22] Immunophenotype C-ALL or pre-B ALL are found with both breakpoints. Conversion from Ph+ALL to CGL chronic phase which occurs in a proportion of cases, is more frequent in *M-BCR* cases,[44,45] but there is evidence that *m-BCR* cases may also convert from ALL to AML which indicates that they have arisen in a pluripotent stem cell.[46]

Target Cell for Ph+ALL

The major distinction between CGL and ALL is the level of differentiation of the target cell for chromosomal change. CGL arises in a pluripotent stem cell capable of differentiating into all myeloid lineages as well to B precursor cells on conversion to the acute or blastic phase. By contrast the target cell for ALL is presumed to be a lymphoid committed progenitor.[47] Evidence from cell colony growth and cell separation studies has shown heterogeneity of lineage involvement in both *m-BCR* and *M-BCR* Ph+ALL. Approximately half the *M-BCR* cases and a minority of *m-BCR* cases investigated have been shown to arise in a pluripotent stem cell suggesting affinity with CGL. The remaining cases involve only cells of the lymphoid lineage indicative of true ALL.[48-50]

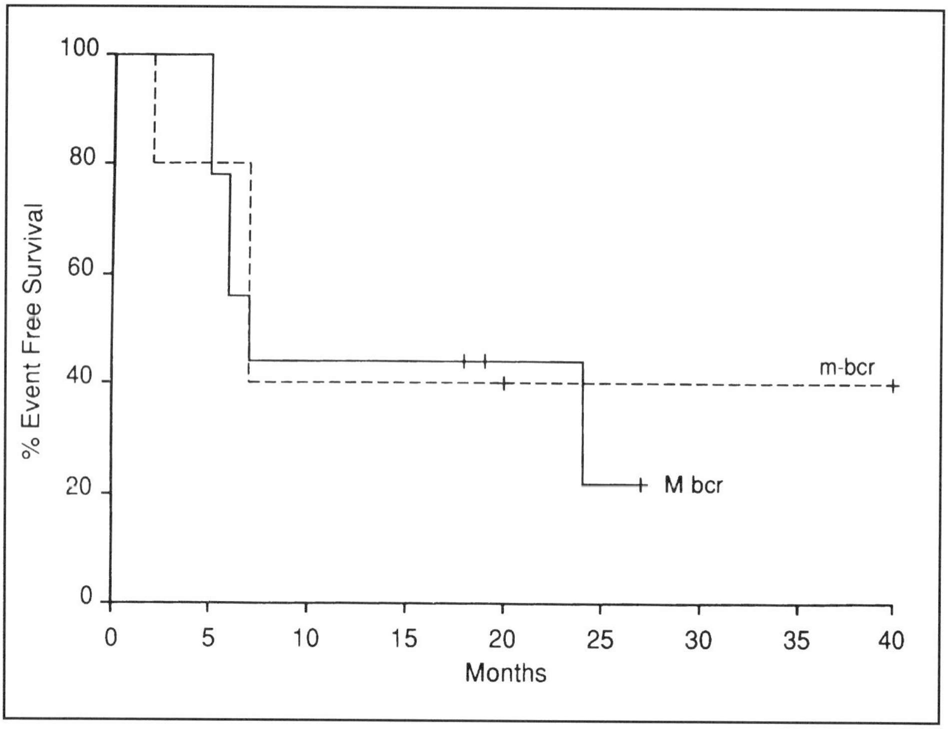

Fig. 5.4. Event-free survival of 20 patients with Ph+ALL with a BCR breakpoint in either the minor (m-bcr) or major (M-bcr) breakpoint cluster region.

Impact of Breakpoint on Prognosis

Prognosis is unrelated to the location of the breakpoint *M-BCR* or *m-BCR* (see Fig. 5.4).[22,49] There is, however, a suggestion that event-free survival may be longer in cases which arise in a pluripotent stem cell and more patients may convert to chronic phase CGL than in cases arising in a lymphoid committed progenitor regardless of breakpoint location (Fig. 5.5).[49]

Treatment for Ph+ve ALL

The poor response of Ph+ve patients to chemotherapeutic regimens has led to the use of bone marrow transplantation in this type of leukemia in particular. The success of this approach is not entirely clear. Some benefit of BMT was reported by Crist et al with 3/7 children who received an allograft remaining disease-free at 10+ to 29+ months.[17] At the same time, extensively reinforced early chemotherapy followed by treatment with pairs of non

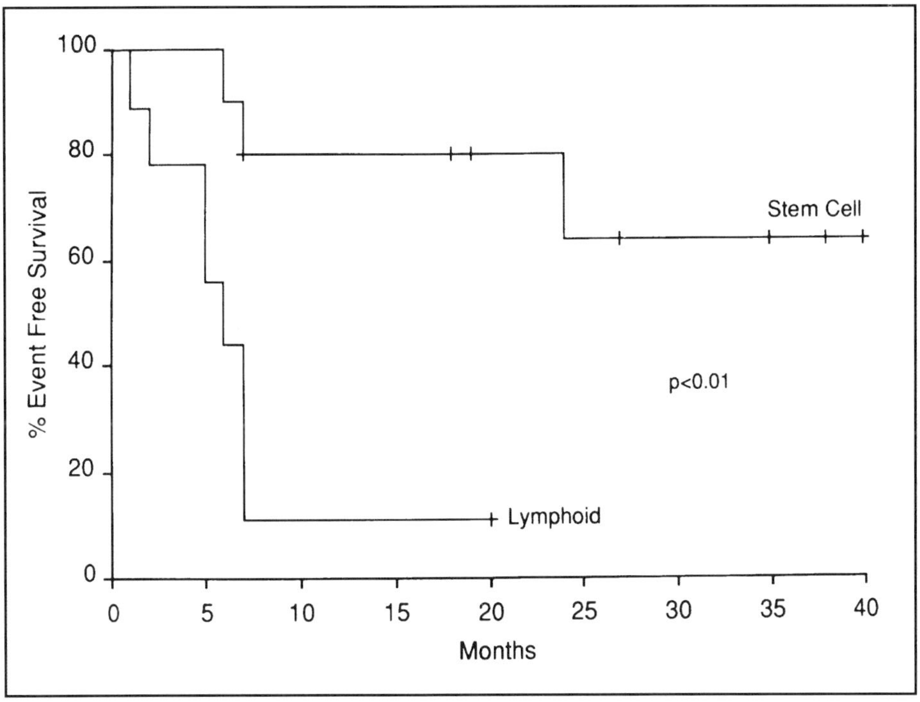

Fig. 5.5. Event-free survival of the same 20 patients as in Figure 5.4 in which the target cell is a pluripotent stem cell (9 cases) or a lymphoid committed progenitor cell (11 cases). Reprinted from Secker-Walker and Craig.[49]

cross-resistant drugs has resulted in a median duration of disease-free survival of 6-8 years in three adolescents and one child with Ph+ALL.[52] BMT for adults with Ph+ve ALL has a variable outcome.[15,53] The results of HLA-identical sibling bone marrow transplants for Ph+ve ALL reported to the International Bone Marrow Transplant Registry identified 21/67 (31%) in continuous complete remission for more than two years after transplant indicating that BMT may be of some value in treating this kind of leukemia.[53]

SUMMARY

In summary, the Philadelphia translocation, (Ph) t(9;22)(q34;q11), identifies an important subgroup in ALL which increases in incidence with age at diagnosis. It is associated with a poor prognosis at all ages. Knowledge of the genetic consequences of the translocation has been used to screen diagnostic and remission samples for t(9;22). The fusion of the genes *BCR* and *ABL* can also be seen on metaphase chromosomes using FISH. FISH has demonstrated hidden Ph rearrangements in chromosomally normal cells. The target cell for chromosomal change may be either a pluripotent stem cell or a cell committed to lymphoid development.

References

1. Nowell PC, Hungerford DA. A minute chromosome in human granulocytic leukemia. Science 1960; 132:1497.
2. Rowley JD. Letter: A new consistent chromosomal abnormality in chronic myelogenous leukaemia identified by quinacrine fluorescence and Giemsa staining. Nature 1973; 243:290-293.
3. Richman CM, Rowley JD, Golomb HM. Chronic granulocytic leukemia: epidemiology and etiology, pathogenesis, cytogenetics, in vitro studies and clinical features. In: Goldman JM Preisler HD, eds. Hematology Leukemias. London: Butterworths International Medical Reviews, 1984: 208-224.
4. Hagemeijer A. Chromosome abnormalities in Chronic myeloid leukemia. Baillieres Clin Haematol 1987; 1:963-981.
5. Propp S, Lizzi FA. Philadelphia chromosome in acute lymphocytic leukemia. Blood 1970; 36:353-360.
6. Biervliet JP, Van Hemel J, Geurts K et al. Philadelphia chromosome in acute lymphocytic leukaemia. Lancet 1975; ii:617.
7. Schmidt R, Dar H, Santorineou M et al. Letter: Ph-1 chromosome and loss and reappearance of the Y chromosome in acute lymphocytic leukaemia. Lancet 1975; 1:1145.
8. Secker-Walker LM, Hardy JD. Philadelphia chromosome in acute leukemia: case report. Cancer 1976; 38:1619-1624.
9. Chessells JM, Janossy G, Lawler SD et al. The Ph1 chromosome in childhood leukaemia. Br J Haematol 1979; 41:25-41.
10. Secker-Walker LM, Summersgill BM, Swansbury GJ et al. Philadelphia-positive blast crisis masquerading as acute lymphoblastic leukaemia in children. Lancet 1976; 2:1405.
11. Beard MD, Durrant J, Catovsky D et al. Blast crisis of chronic myeloid leukaemia (CML). I. Presentation simulating acute lymphoid leukaemia (ALL). Br J Haematol 1976; 34:167-178.
12. Bloomfield CD, Peterson LC, Yunis JJ et al. The Philadelphia chromosome (Ph1) in adults presenting with acute leukaemia: a comparison of Ph1+ and Ph1-patients. Br J Haematol 1977; 36:347-358.
13. Catovsky D. Ph1-positive acute leukaemia and chronic granulocytic leukaemia: one or two diseases? Br J Haematol 1979; 42:493-498.
14. Tuszynski A, Dhut S, Young BD et al. Detection and significance of bcr-abl mRNA transcripts and fusion proteins in Philadelphia-

positive adult acute lymphoblastic leukemia. Leukemia 1993; 7:1504-1508.

15. Annino L, Ferrari A, Cedrone M et al. Adult Philadelphia-chromosome-positive acute lymphoblastic leukemia: experience of treatments during a ten-year period. Leukemia 1994; 8:664-667.
16. IWCL3. Third International Workshop on Chromosomes in Leukemia (1980): Chromosomal abnormalities in acute lymphoblastic leukemia: Structural and numerical changes in 234 cases. Cancer Genet Cytogenet 1981; 4:101-110.
17. Crist W, Carroll A, Shuster J et al. Philadelphia chromosome positive childhood acute lymphoblastic leukemia: clinical and cytogenetic characteristics and treatment outcome. A Pediatric Oncology Group study. Blood 1990; 76:489-494.
18. Gotz G, Weh HJ, Walter TA et al. Clinical and prognostic significance of the Philadelphia chromosome in adult patients with acute lymphoblastic leukemia. Ann Hematol 1992; 64:97-100.
19. Secker-Walker LM, Prentice HG, Durrant J et al. Cytogenetics adds independent prognostic information in adults with acute lymphoblastic leukaemia on MRC trial UKALL XA. Br J Haematol 1997; in press.
20. Ribeiro RC, Abromowitch M, Raimondi SC et al. Clinical and biologic hallmarks of the Philadelphia chromosome in childhood acute lymphoblastic leukemia. Blood 1987; 70:948-953.
21. Sessarego M, Defferrari R, Fugazza G et al. Involvement of the short arm of the derivative chromosome 9 in Philadelphia-positive acute lymphoblastic leukemia. Cancer Genet Cytogenet 1991; 52:43-49.
22. Secker-Walker LM, Craig JM, Hawkins JM et al. Philadelphia positive acute lymphoblastic leukemia in adults: age distribution, BCR breakpoint and prognostic significance. Leukemia 1991; 5:196-199.
23. Russo C, Carroll A, Kohler S et al. Philadelphia chromosome and monosomy 7 in childhood acute lymphoblastic leukemia: a Pediatric Oncology Group study. Blood 1991; 77:1050-1056.
24. Kurzrock R, Talpaz M. The genetics of the Philadelphia chromosome. In: Burnett A, Newland A Keating A, eds. Haematological Oncology Volume 1. Cambridge University Press, 1991: 79-110.
25. IWCL3. Third International Workshop on Chromosomes in Leukemia (1980): Clinical significance of chromosomal abnormalities in acute lymphoblastic leukemia. Cancer Genet Cytogenet 1981; 4:111-137.
26. Radich JP, Kopecky KJ, Boldt DH et al. Detection of BCR-ABL fusion genes in adult acute lymphoblastic leukemia by the polymerase chain reaction. Leukemia 1994; 8:1688-1695.

52. Roberts WM, Rivera GK, Raimondi SC et al. Intensive chemotherapy for Philadelphia-chromosome-positive acute lymphoblastic leukaemia. Lancet 1994; 343:331-332.
53. Barrett AJ, Horowitz MM, Ash RC et al. Bone marrow transplantation for Philadelphia chromosome-positive acute lymphoblastic leukemia. Blood 1992; 79:3067-3070.
54. Secker-Walker LM. Prognostic and biological importance of chromosome findings in acute lymphoblastic leukemia. 10th anniversary article. Cancer Genet Cytogenet 1990; 49:1 13.

CHAPTER 6

Mature B Cell and T ALL

Many chromosomal translocations seen in patients with mature B cell ALL or with T ALL have a breakpoint at the same locus as one of the immunoglobulin or T cell receptor genes, respectively. These occur in more than 80% of B cell patients and in up to 45% of T cell patients. The molecular genetics of these translocations were the earliest to be investigated and these investigations led to the demonstration of one of the ways in which chromosomal abnormalities may bring about the activation of an oncogene. There are other abnormalities associated with B and T ALLs but they are not specific to patients with these particular immunophenotypes. The cytogenetics and genetics of mature B and T ALL will be described in this chapter.

MATURE B CELL ALL

t(8;14)(q24;q32),t(2;8)(p12;q24) AND t(8;22)(q24;q11)

Burkitt's type ALL, FAB type L3, represents an acute leukemic presentation of Burkitt's lymphoma.[1,2] This type of acute leukemia accounts for approximately 5% of all ALLs and is associated with three classic translocations t(8;14) and its variants t(2;8) and t(8;22). Diagrams of these translocations are shown in Figures 6.1, 6.2 and 6.3. A full and partial karyogram with t(8;14)(q24;q32) are shown in Figures 6.4 and 6.5. These translocations are found in Burkitt's lymphoma and also in mature B ALL.[1,3,4] Until recently, they were believed not to occur in B-precursor ALL but a few reports of these translocations in earlier B lineage ALL indicate that there may be exceptions to this rule. Almost all cases with these translocations are of L3 morphology. Again, there are rare cases with L1 or L2 blasts.

Reviews by Berger and Bernheim in 1985[5] and Kornblau et al in 1991[6] of 122 and 170 cases, respectively, of Burkitt's lymphoma

(BL) or Burkitt's -type leukemia (ALL L3) indicated that within the spectrum of this type of mature B cell malignancy over 80% had one of the classic translocations. Sixty-two percent of patients either had t(8;14) or, in earlier publications, were described as having abnormalities of 14q+ or 8q-. These abnormalities were presumed to be examples of t(8;14) which could not be fully analyzed. Twelve percent of patients had t(8;22) and 9% had t(2;8). This left just 17% with other abnormalities or in whom no abnormality was detected. Since 1983, there have been reports of at least 22 patients with Burkitt's-type disease and AIDS. Six of the 22 AIDS patients have had t(8;22) which makes this translocation significantly more frequent in AIDS patients than in others.[6] Additional chromosomal change is found at diagnosis in well over half the reported cases with a predominance of trisomy 1q and trisomy 7, 8, 12 and 18 and structural change involving 6q11-14 and 17p.[4-8] An example of a complex clone from a patient with B-ALL is shown in Figure 6.6.

Genetic Change

The translocations have in common a breakpoint at 8q24, site of the oncogene *MYC*, and a second breakpoint at the site of an immunoglobulin chain gene, or the heavy chain *IGH* at 14q32, or the light chain genes kappa *IGK* at 2p12, or lambda *IGL* at 22q11 (see Table 6.1). The *MYC* oncogene has three exons of which only exons 2 and 3 encode for proteins.[9,10]

In t(8;14), the chromosome 8 breakpoint is always 5'of the second *MYC* exon which means the entire coding region is trans-

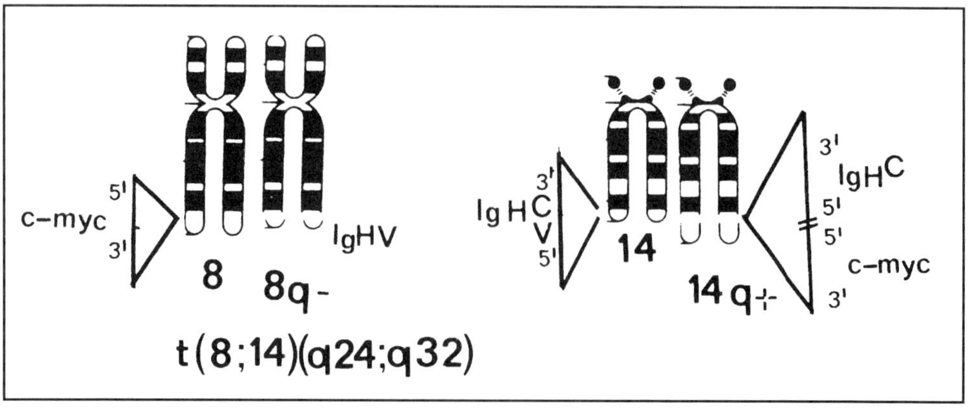

Fig. 6.1. Ideogram to show t(8;14)(q24;q32).

Table 6.1. Translocations and genes in mature B ALL

Type of gene and translocation	Affected gene	Rearranging gene
Basic-helix-loop-helix		
t(8;14)(q24;q32)	c-*MYC* (8q24)	*1GH* (14q32)
t(2;8) (p12;q24)	c-*MYC* (8q24)	*1GK* (2p12)
t(8;22) (q24;q11)	c-*MYC* (8q24)	*1GL* (22q11)

located to 14q32. The 8q24 breakpoints show considerable variability and may occur over an area of 170-190 kb.[11] The break may occur upstream of the entire gene or may lie within the gene in exon one or in the first intron. The 14q32 breakpoint occurs within *IGH* in the switch (S) or joining (J) regions. It may occasionally involve the variable (V) region or a constant (C) region coding segment. The orientation of *MYC* and *1GH* on the derived chromosome 14 is 5' to 5' (or head to head) (see Fig. 6.7).

Different clusters of breakpoints have been observed for patients with Burkitt's lymphoma and t(8;14) in different parts of the world suggesting a role for geographical or genetic factors. In endemic African cases (eBL), the breakpoints lie more than 100 kb upstream or 5' of *MYC,* while those on 14 are located within the *IG* joining region *(JH)*, the region involved in rearrangement early in B cell development. In the sporadic Caucasian cases (sBL), on the other hand, the breakpoints on *MYC* are often within the 5' portion (exon1/intron1) of the *MYC* gene or in its 5' flanking sequences. Moreover, chromosome 14 breaks in the switch region of the constant μ (Cμ) region of *IGH* which is involved relatively late in B-cell differentiation.[12] Further evidence for geographical variation in breakpoints is provided by the investigation by Saglio et al of 15 Spanish and Italian patients with t(8;14). Two main breakpoints on 14 were seen: in the *JH* region (6 cases) or in the C alpha 1 region (6 cases). This latter, accounting for 40% of their patients, has only occasionally been described by others.[13]

In t(2;8) and t(8;22) the chromosome 8 breakpoint lies more than 20 kb 3' of *MYC,* leaving the structure of *MYC* intact on

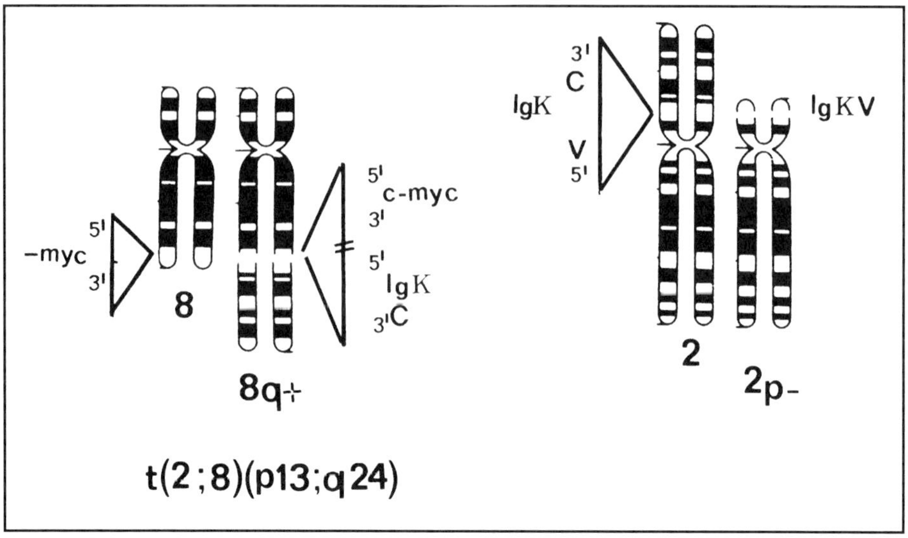

Fig. 6.2. Ideogram to show t(2;8)(p13;q24).

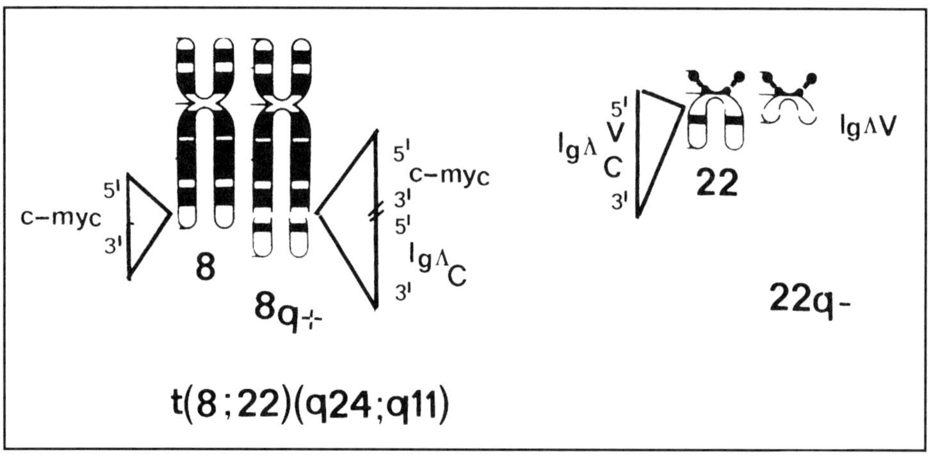

Fig. 6.3. Ideogram to show t(8;22)(q24;q11).

the rearranged chromosome 8.[14] The breakpoints on 2p and 22q occur within the kappa and lambda light chain gene, respectively. Orientation of the genes following the translocation in t(2;8) and t(8;22) is 5' to 3' or head to tail.

The recombination of *MYC* and the *IGH* genes appears to represent misrecognition of the *MYC* sequences by a recombinase involved in *IGH* gene rearrangement.[15] The result of the translocations is to downregulate *MYC* expression. This is brought about by the proximity of *MYC* to the immunoglobulin gene but not necessarily as a result of gene fusion. Overexpression of *MYC* brings about uncontrolled cell proliferation.[9] The c-*MYC* protein has several functional domains including a basic region for DNA binding and helix-loop-helix (b-HLH) and leucine zipper (ZIP) protein dimerization motifs. Activation of *MYC* by translocation compromises a transcriptional network involving at least three other factors all of which also have b-HLH and ZIP domains. *MYC* can dimerize with a protein called *MAX* which can bind to DNA and to proteins MAD and Mxi-1. All these proteins are presumably in monomer-dimer equilibrium in the normal cell. c-*MYC-MAX* dimers are transcriptionally active where-as *MAX-MAD* or *MAX-Mxi-1* dimers are not. After translocation, the equilibrium is disrupted by incorrect expression of c*MYC*. As the c-*MYC-MAX* dimer is necessary for oncogene activation, excess c-*MYC* expression would be expected to cause a shift in equilibrium and presumably result in transcription of downstream target genes leading to oncogenesis.[10]

Clinical Features

Burkitt's-type disease has been described in patients at all ages from 3-80 years old of which 63% have been aged <30 years.[6] Seventy-five percent of patients are male. Blasts are of FAB type L3 and of mature B cell type expressing surface immunoglobulin (sIg). White blood cell counts are not usually raised (median white blood cell count is 10 x 10^9/L) but extra medullary disease, particularly CNS disease, is frequent. One study has shown no statistically significant difference in age or sex ratios between patients with mature B cell ALL having one of the classic translocations (with an 8q24 breakpoint) and those without 8q24 abnormalities. All these patients had L3 blasts although the proportion of L3 blasts in some cases was low, raising the possibility that accounts of B-ALL with blasts of other FAB types may have been exaggerated.[16]

Prognosis

For many years patients with these translocations were treated with ALL protocols and prognosis for this subgroup was extremely

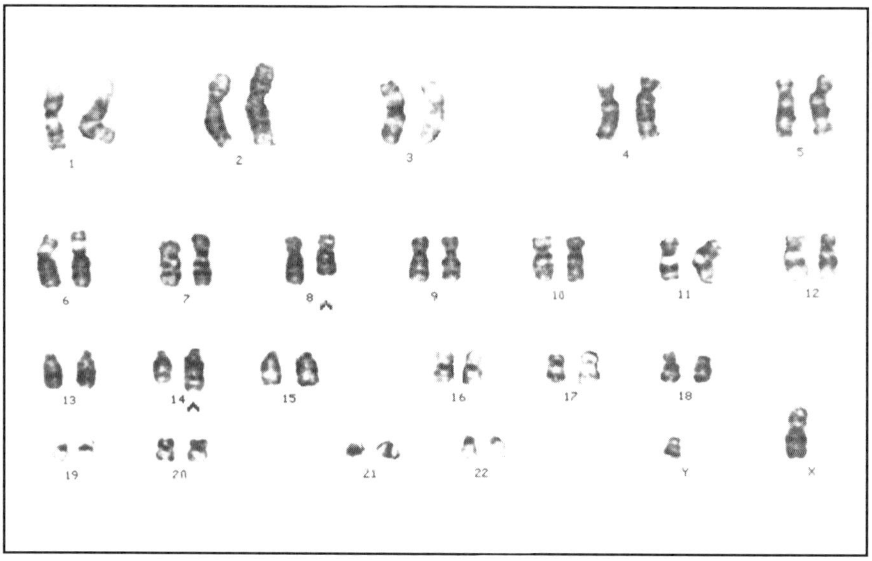

Fig. 6.4. Karyogram with 46,XY,t(8;14)(q24;q32).

poor. Half the cases failed to achieve remission and those that did relapsed early. Median survival reported for these cases was <1 year,[6,17] and increasing age did not adversely affect survival.[6] In one survey, remission rates (43%) and median survival (4.8 months) of patients with one of the classic translocations compared unfavorably with those of patients without 8q24 abnormalities (68% and 18.4 months, respectively).[16] A larger review showed that patients with variant translocations did less well than those with the classic t(8;14)(q24;q32), but the difference in survival between the two groups was not statistically significant.[6] Pooled survival data must be regarded cautiously because of the numerous treatment regimens which have been used. However, recently patients with B-ALL have been given short-term intensive treatment as for non-Hodgkin's lymphoma and this has resulted in considerably improved survival over those treated on ALL regimens. These patients are at high risk of CNS relapse, but prevention can be achieved in most patients with the use of high-dose intravenous and intrathecal chemotherapy without cranial irradiation.[18] The impact of karyotype on survival in patients on these new protocols will be of interest.

t(14;18)(q32;q21) IN ALL

This translocation traditionally associated with mature B cell lymphomas has occasionally been described in cases of C-ALL. In three cases the translocation has been found in association with t(8;14) or del(9)(p). These could be explained by the coexistence of both a low-grade follicular lymphoma (with t(14;18)) and BL (with t(8;14)), as has been described in one case.[19] Lack of sIg in such cases may result from the involvement of the *IGH* genes on both chromosomes in a translocation. Association of t(14;18) and t(8;9)(q24;p13) has been described in three cases with de novo ALL[20,21] and association of t(14;18) with t(1;19) has been described both in C-ALL[22] and in mature B cell ALL.[23]

T ALL

Cytogenetics

The majority of translocations in T ALL involve the chromosomal sites of the T cell receptor genes at 14q11, 7q35 or 7p15. The reported incidence of translocations involving these sites ranges from 21% to 42% of patients (see Table 6.2).[24-27] The involvement of *TCR* loci in T ALL, found in approximately 30% of patients is much lower than the involvement of the *IG* loci in mature B ALL.

Other chromosomal abnormalities found in T ALL but not specific to this immunophenotype are del(6)(q), and abnormalities of 9p and of 12p. These may occur alone or as part of a complex clone which may include one of the translocations listed above.[24,26-28]

Table 6.2. Incidence of translocations involving T-cell receptor sites in T ALL

Total cases	TCR+	Refs
57	12	Raimondi et al[27]
50	20	Kaneko et al[25]
44	12	Berger et al[24]
19	8	Secker-Walker et al[26]
170	52 (31%)	Total

Table 6.3. Translocations and genes in T ALL

Type of gene and translocation	Affected gene	Rearranging gene
Basic-helix-loop-helix		
t(8;14)(q24;q11)	c-*MYC*(8q24)	*TCRA*
t(1;14)(q32:q11)	*TAL1/SCL* (1p32)	*TCRA*
t(1;7)(p32;q35)	*TAL1/SCL* (1p32)	*TCRB*
t(7;19)(q35;p13)	*LYLI* (19p13)	*TCRB*
t(7;9)(q35;q34)	*TAL2* (9q34)	*TCRB*
LIM proteins		
t(11;14)(p15;q11)	*RBTNI* (11p15)	*TCRD*
t(11;14)(p13;q11)	*RBTN2* (11p13)	*TCRD/A*
Homeobox protein		
t(10;14)(q24;q11)	*HOX 11* (10q24)	*TCRA/D*
Other		
Inv(14(q11q32)	*TCL1* (14q32.1)	*TCRA*
t(7;9)(q34;q34.3)	*TANI* (9q34.3)	*TCRB*
t(1;7)(p34;q34)	*LCK* (1p34)	*TCRB*
t(5;14)(q33-34;q11)		*TCRD*

The translocation sites frequently involved in T ALL have in common the location of one of the T-cell receptor (*TCR*) genes. The alpha and delta genes of the *TCR* -*TCRA* and *TCRD* have been mapped to 14q11-q13 and the beta and gamma chains *TCRB* and *TCRG,* respectively to 7q32-36 and 7p15, respectively (see Table 6.3).

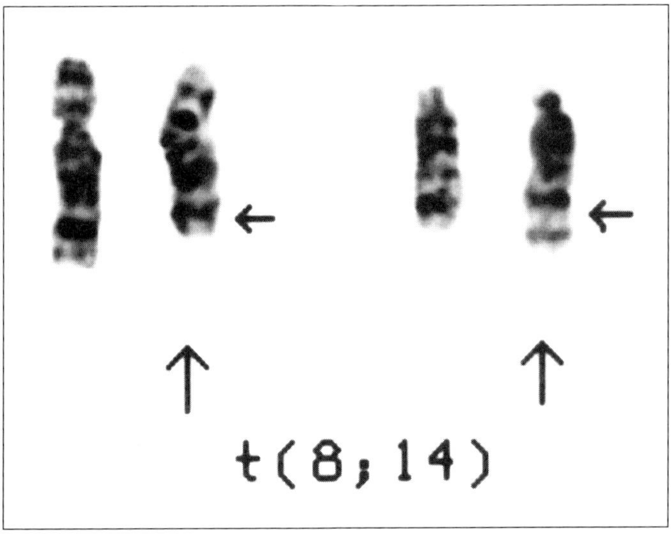

Fig. 6.5. Partial karyogram to show t(8;14)(q24;q32).

Clinical Features

T ALL shares many clinical features with lymphoblastic lymphoma. These disorders are more frequent in males than females with a ratio of approximately 4:1 and the median age of childhood patients is nine years. An anterior mediastinal mass is seen on chest X-rays in 50% of cases and median WBC is 50 x 10^9/L. Central nervous system disease occurs more frequently than in C-ALL. Detailed immunophenotyping shows three stages of blast cell maturity: early thymocyte, common thymocyte and late thymocyte.[24,29]

Prognosis

Response to therapy, achievement of complete remission and survival is less good than that for C-ALL but risk-directed therapy has generally improved survival for patients with T ALL.

t(1;14)(p32-34;q11)

t(1;14) accounts for between 3% and 6% of T ALL cases.[30,31] Additional change which is present in approximately 50% of cases, frequently includes partial deletion of 6q.[30,31]

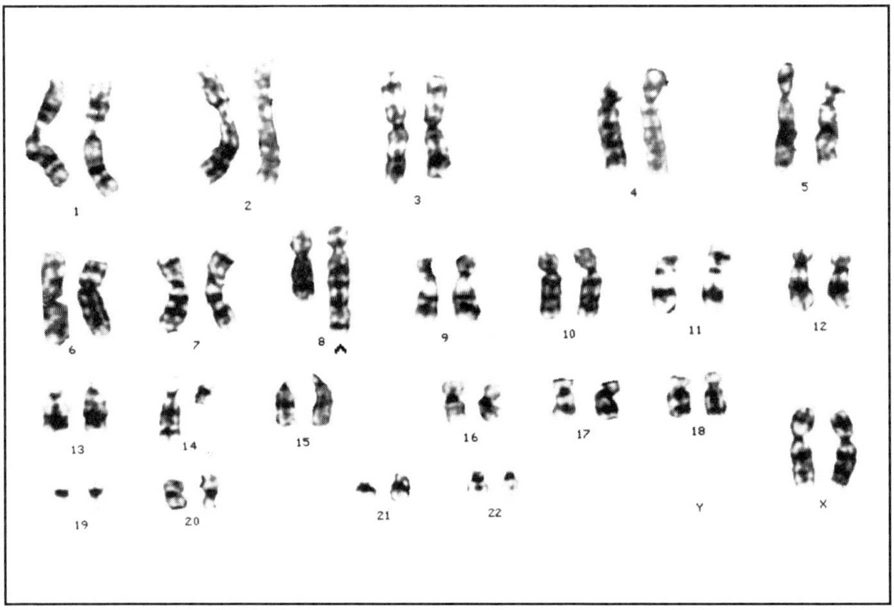

Fig. 6.8. Karyogram with 46, XX, t(8;14)(q24;q11).

The consequence of both alterations is overexpression of the *SCL/TAL1* gene in leukemic cells. However, causal involvement of *SCL/TAL1* in neoplastic transformation has not been demonstrated. Transgenic mice, expressing the *TAL1* transgene at high levels, showed no alterations in T cell development and these mice did not develop T cell tumors.[37]

Incidence of *TAL1* involvement at diagnosis

Involvement of the *TAL1* gene which occurs only in T ALL is found in up to 30% of patients, a considerably higher incidence than that demonstrated cytogenetically for the involvement of 1p32.[32] A *TAL1* lesion has been demonstrated in 7/76 (9%) of children and in 2/66 (3%) of adults with T ALL.[32-36] Analysis of 313 acute leukemias by Southern blotting, using a specific *TAL1* deletion probe, suggested that these deletions occurred exclusively in T ALL and did not occur in precursor B-ALL, in AML or in normal blood cells or thymocytes.[38] Two other translocations which disrupt the *HTAL1* gene are t(1;3)(p34;p21)[40] and t(1;7)(p32;q35), which latter juxtaposes *TAL1* to *TCRB* in T ALL.[41]

MRD detection

Janssen et al have demonstrated that PCR, using *TAL1* fusion region-specific probes, may be used for the detection of minimal residual disease for patients with t(1;14) and that this technique has a sensitivity of 10^{-5}.[36,38,39] The significance of persistence of the clone at this level is as yet not known.

t(5;14)

t(5;14)(q33-34;q11) has been described in four children, two with T ALL and two with early B lineage, and is associated with rearrangement of *TCRD*. The patients with T ALL were aged 14 months, 2 years, 10 years and 12 years and each presented with one or more features of bulky disease. Survival for the two patients with B-lineage ALL was more than 19 and 22 months, while the two T cell patients fared poorly. One patient relapsed six months from diagnosis and the other developed acute myeloid leukemia 17 months from diagnosis but still with t(5;14).[42]

t(8;14)(q24;q11)

A karyogram showing this translocation is shown in Figure 6.8. A survey of 15 childhood cases, described as having pediatric

Table 6.4. 6q abnormalities

Total Cases	abn6q	T ALL	B-lin	Refs
412	45	10/72	35/340	Hayashi et al[5]
135	11	3/27	8/100	Vanderplas et al[63]
411	26	7/56	17/334	GFCH[64]
392	42	9/57	33/335	Raimondi et al[27]
1186	82	9/110	49/753	Appendix 2
2536	206	38/322	142/1862	Total
	(8%)	11.8%	7.6%	

Table 6.5. 9p abnormalities in childhood ALL

Total cases	abn 9p	T ALL abn9p/Total	B-lin ALL abn9p/Total	Refs
135	16	4/27	11/100	Vanderplas et al[63]
398	40	10/61	30/337	Murphy et al[74]
411	40	7/56	29/334	GFCH[30]
392	41	8/57	33/335	Raimondi et al[27]
1186	67	10/110	42/753	Appendix 2
2522	204	39/311	145/1859	Total
	8%	12.5%	7.8%	

leukemia/lymphoma with this translocation, gave an estimated incidence of 1% of cases.

The t(8;14)(q24;q11) brings *TCRA* into juxtaposition with *MYC* and brings about the downregulation of this oncogene. This mechanism is analogous to that of t(8;14)(q24;q32) in B cell malignancies in which *C-MYC* is downregulated by the *IGH* gene.[43,44]

Clinical features

The disease is characterized by male predominance (2:1), a median age of 5.5 years (range 1.8-17 years), high white blood cell count (median 95 x 10^9/l), central nervous system involvement and extra medullary leukemia. A T cell immunophenotype has been most commonly found, but at least one patient with B lineage precursor ALL has been described. The same translocation has been seen in two Arab children from the Gaza strip; one of whom had a brother who died of T cell ALL a few years later.[45] Such cases may provide valuable clues to the role of genotype in the development of ALL. Prognosis has been poor with a median event-free survival of four months.[43]

t(10;14)(q24;q11).

An example of the t(10;14) translocation is shown in Figure 6.9. This translocation was described as a recurring abnormality in

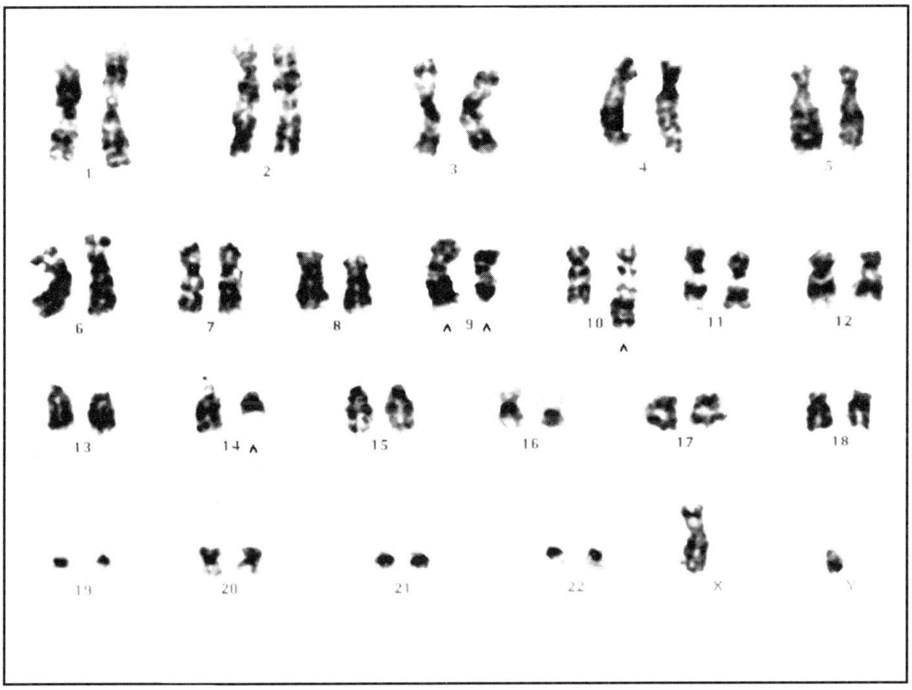

Fig. 6.9. Karyogram with 46, XX, t(10;14)(q24;q11).

T ALL in 1986.[46] It has since been described in 5-10% of cases of T ALL. Molecular analyses have revealed involvement of *TCRD* and a gene located at 10q24 described as *TCL-3,* which has been renamed *HOX11* consistent with its role as a homeobox gene.[47-49] This gene is expressed in liver but has not been detected in normal thymus or T cells. Its translocation into the T cell receptor locus is presumed to bring about deregulation of *HOX11* and fulfillment of its oncogenic potential.[49,50] Immortalization of hematopoietic precursor cells by *HOX11* has been demonstrated in mice.[51]

t(11;14)(p15;q11) AND **t(11;14)(p13;q11).**

These translocations occur in both T ALL and lymphoblastic lymphoma (LBL). The t(11;14) (p15;q11) has been shown to account for 1/103 and 2/50, between 1% and 4%, of T cell cases.[25,52] However, t(11;14)(p13;q11), with a more proximal breakpoint on 11p (see Figure 6.10), appears to be slightly more frequent, accounting for 7/103 and 3/50, or approximately 7% of cases, in

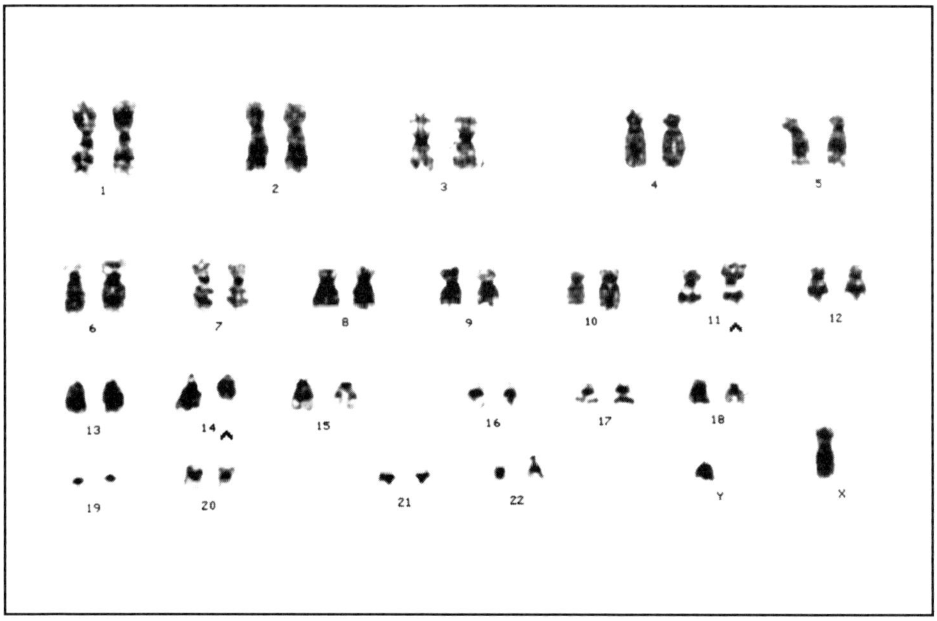

Fig. 6.10. Karyogram with 46, XY, t(11;14)(p13;q11).

the same series of T ALL.[25,52] The genes located at 11p15 and 11p13 are *RBTN1* and *RBTN2,* respectively. These genes have been shown to be juxtaposed to the *TCRD* locus at 14q11. *RBTN1* and *RBTN2* have been shown to induce T cell tumors in transgenic mice, but are also essential for erythroid differentiation in normal mouse erythropoiesis.[51,53]

Clinical features

The clinical features of these cases are typical of T ALL with male predominance (6/7 cases), age >10 years, WBC >100 x 10^9/L, presence of mediastinal mass and nervous system involvement. Immunophenotypes showed more cases with CD4 and CD8 doubles than expected (6/6 v 35/86 cases tested), indicative of intermediate or late stage thymocyte differentiation.[52]

inv(14)(q11q32) is a common aberration in T cell chronic malignancies but is rare in T ALL. Only a few cases have been described.[27,54,55]

TRANSLOCATIONS OF 7q32-q36

t(1;7)(p34;q34)

This translocation has been shown to result in the fusion of the *TCRB* constant region and transcriptional enhancer with the *LCK* gene. *LCK* is normally expressed from two distinct promoters. A proximal promoter initiates type I transcripts and is primarily expressed in thymocytes. A distal promoter initiates type II transcripts and is primarily expressed in mature thymocytes. Both transcripts encode the same protein kinase. Two cell lines each with t(1;7) have been investigated and shown to have different breakpoints on chromosome 1. One of these resulted in translocation of the type II promoter to the derivative 7 chromosome. The other break occurred upstream of the type II promoter, leaving an intact *LAC* gene on the derived chromosome 1. Transcripts of both types I and II were found in each case leading the authors to conclude that type II transcripts may originate from the normal *LCK* allele.[56]

t(7;9)(q34;q32)

The *TAL2* gene located at 9q32 produces a helix-loop-helix protein dimerization and DNA binding domain, homologous to those encoded by *TAL1* and *LYL1*. It thus makes the third in a family of proteins of importance in T cell leukemogenesis. In one investigated case with t(7;9), investigated the *TAL2* gene was found 33 kb from the chromosome 9 breakpoint juxtaposed to *TCRB*. Active transcription of *TAL2* sequences have been shown in a cell line with t(7;9).[57]

t(7;11)(q35;p13)

This translocation probably accounts for about 2% of T ALLs.[25] Involvement of *RBTN2* in T ALL cases with this translocation has been investigated. Using a very sensitive PCR-based assay, a trace expression of *RBTN2* was detected at a level of 0.01% of the *TCRB* message. *RBTN2* expression was not seen in normal thymocytes or on patients with abnormality of 11p13. Expression of *RBTN2* was demonstrated in T ALLs in the presence of the translocation

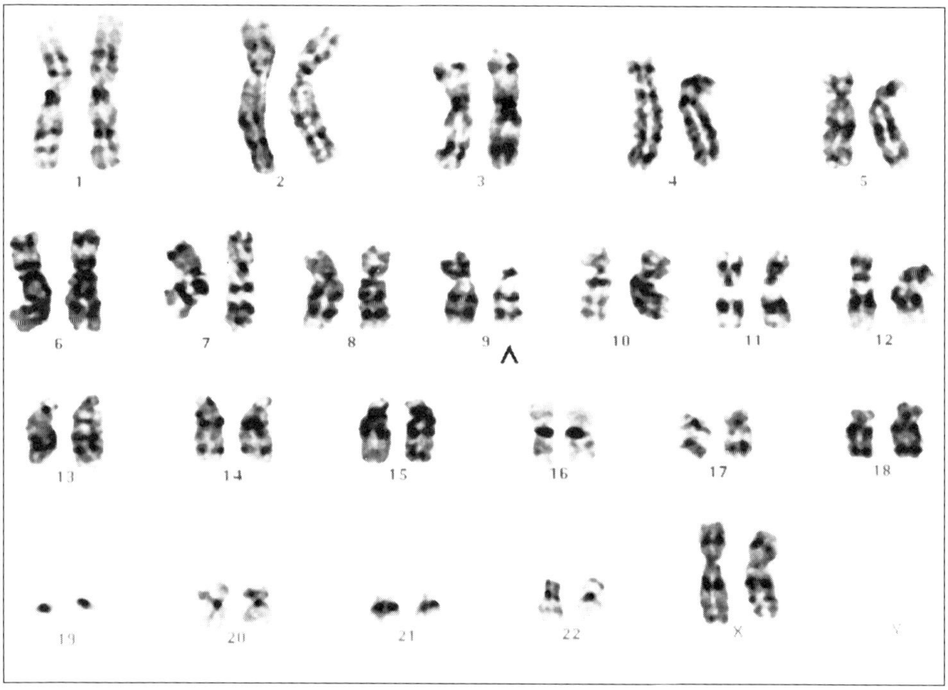

Fig. 6.12. Karyogram with 46, XX, del(9)(p13p24).

cyclin D-dependent serine-threonine kinases, CDK4 and CDK6. These are specifically activated during G1 phase and are necessary for the transition from G1 to S-phase. Analysis of 43 pediatric patients with interphase FISH, Southern blot analysis and RT-PCR identified deletions of these genes in 18/20 patients with chromosomal abnormalities of 9p and in 5/23 patients with apparently normal chromosomes 9p. This confirms deletion of these loci as among the more frequent in ALL in childhood.[67] Homozygous loss of $p16^{INK4}$ has been reported in 4/14 ALLs,[68] in 1/31 B lineage ALLs and in 20/24 T ALLs.[69] Recent evidence suggests that homozygous loss of this gene may occur in >80% of T ALLs. Homozygous deletion in adults is much lower.

Clinical associations

9p abnormalities were originally described in patients with features characteristic of T ALL, i.e., massive lymphadenopathy, splenomegaly, CNS disease and a high peripheral leukocyte count.[70,71] Subsequently, 9p loss has been found to occur also in patients without lymphomatous disease.[28,72-74] A compilation of recent studies shown in Table 6.5 shows that abn9p is found significantly more often in T ALL (12.5%) than in early B-ALL (7.8%) ($p < 0.01$). The impact of 9p involvement on prognosis is unclear.[73,74]

Isochromosome 9q

Isochromosome 9q, i(9)(q10q10) results in the loss of the short arm of one homolog of chromosome 9 accompanied by duplication of the long arm.[73] Cases reported range from 3-35 years, WBC from 1.4->250 x 10^9/L and this abnormality has been seen in cases with blast cell immunophenotypes C-, Pre-B or T ALL. It is probably more common among childhood than adult cases; there is no evidence that it is associated with a poor prognosis.[75] This will be discussed with other isochromosomes more fully in chapter 6.

Abnormalities of Chromosome 12p

Abnormalities of 12p have been described in between 5% and 18% of cases of T ALL.[24,26-28] A full description of the translocations and deletions of 12p together with the genetic consequences of 12p involvement are covered in chapter 7.

Normal Cytogenetics in T ALL

The incidence of cytogenetically normal cases in T ALL is consistently reported to be higher than that in C-ALL and ranges between 20% and 28%.[24-27] There is some evidence that chromosomally normal cases are found particularly among those with immunophenotypes of the least mature thymocytes.[26,27] The possibility that a number of these patients have submicroscopic genetic rearrangements must be considered. Pui et al have shown that T cell patients with normal karyotypes had a better prognosis than those with an abnormal clone.[29]

SUMMARY

In summary, the translocations found in >80% of patients with mature B cell ALL are the same as those found in Burkitt's lymphoma (BL): t(8;14), t(2;8) and t(8;22). In BL, the translocations bring about the juxtaposition of the oncogene *MYC* (at 8q24) and one of the immunoglobulin genes *IGH* (at 14q32), *IGK* (at 2p12), or *IGL* (at 22q11). Translocations found in T ALL involve juxtaposition of an oncogene (of which 10 have been described so far), with one of the T cell receptor genes *TCRA/D* (at 14q11), *TCRB* (at 7q35) or *TCRG* (at 7p15). These occur in up to 40% of T ALL cases. In mature B ALL and in T ALL, downregulation of the oncogene is brought about by the juxtaposition of genes rather than by gene fusion. Three other abnormalities found in T ALL but not exclusive to it are: abnormalities of 6q, 9p and 12p.

References

1. Berger R, Bernheim A, Brouet JC et al. t(8;14) translocation in a Burkitt's type of lymphoblastic leukaemia (L3). Br J Haematol 1979; 43:87-90.
2. Mitelman F, Andersson Anvret M, Brandt L et al. Reciprocal 8;14 translocation in EBV-negative B-cell acute lymphocytic leukemia with Burkitt-type cells. Int J Cancer 1979; 24:27-33.
3. Berger R, Bernheim A, Flandrin G et al. [Translocation t(8;14) in acute lymphoblastic leukaemia, Burkitt type (author's transl)]. Nouv Presse Med 1979; 8:181-183.
4. Berger R, Bernheim A. Cytogenetic studies on Burkitt's lymphoma-leukemia. Cancer Genet Cytogenet 1982;7:231-244.
5. Berger R, Bernheim A. Cytogenetics of Burkitt's lymphoma-leukaemia: a review. IARC Sci Publ 1985:65-80.
6. Kornblau SM, Goodacre A Cabanillas F. Chromosomal abnormalities in adult non-endemic Burkitt's lymphoma and leukemia: 22 new reports and a review of 148 cases from the literature. [Review]. Hematological Oncology 1991; 9:63-78.
7. Knuutila S, Elonen E, Heinonen K et al. Chromosome abnormalities in 16 Finnish patients with Burkitt's lymphoma or L3 acute lymphocytic leukemia. Cancer Genet Cytogenet 1984; 13:139-151.
8. Secker-Walker LM, Stewart E, Norton J et al. Multiple chromosome abnormalities in a drug resistant TdT positive B-cell leukemia. Leukemia Research 1987; 11:155-161.
9. Croce CM, Nowell PC. Molecular basis of human B cell neoplasia. Blood 1985; 65:1-7.
10. Rabbitts TH. Chromosomal translocations in human cancer. Nature 1994; 372:143-149.

11. Joos S, Falk MH, Lichter P et al. Variable breakpoints in Burkitt lymphoma cells with chromosomal t(8;14) translocation separate c-myc and the IgH locus up to several hundred kb. Human Molecular Genetics 1992; 1:625-632.
12. Shiramizu B, Magrath I. Localization of breakpoints by polymerase chain reactions in Burkitt's lymphoma with 8;14 translocations. Blood 1990; 75:1848-1852.
13. Saglio G, Grazia Borrello M, Guerrasio A et al. Preferential clustering of chromosomal breakpoints in Burkitt's lymphomas and L3 type acute lymphoblastic leukemias with a t(8; 14) translocation. Genes Chromosom Cancer 1993; 8:1-7.
14. Zeidler R, Joos S, Delecluse HJ et al. Breakpoints of Burkitt's lymphoma t(8;22) translocations map within a distance of 300 kb downstream of MYC. Genes Chromosom Cancer 1994; 9:282-287.
15. Haluska FG, Finver S, Tsujimoto Y et al. The t(8;14) chromosomal translocation occurring in B-cell malignancies results from mistakes in V-D-J joining. Nature 1986; 324:158.
16. Davey FR, Lawrence D, MacCallum J et al. Morphologic characteristics of acute lymphoblastic leukemia (ALL) with abnormalities of chromosome 8, band q24. Am J Hematol 1992; 40:183-191.
17. IWCL3. Third International Workshop on Chromosomes in Leukemia (1980): Clinical significance of chromosomal abnormalities in acute lymphoblastic leukemia. Cancer Genet Cytogenet 1981; 4:111-137.
18. Hoelzer D, Ludwig W-D, Eckhard E et al. Improved outcome in adult B-cell acute lymphoblastic leukemia. Blood 1996; 87:495-508.
19. Kramer MH, Raghoebier S, Beverstock GC et al. De novo acute B-cell leukemia with translocation t(14;18): an entity with a poor prognosis. [Review]. Leukemia 1991; 5:473-478.
20. Lillington DM, Monard S, Johnson PW et al. The t(14;18) in a patient with de novo acute lymphoblastic leukemia is associated with t(8;9). Leukemia 1994; 8:560-563.
21. Nacheva E, Dyer MJ, Fischer P et al. C-MYC translocations in de novo B-cell lineage acute leukemias with t(14;18)(cell lines Karpas 231 and 353). Blood 1993; 82:231-240.
22. Wlodarska I, Stul M, De Wolf-Peeters C et al. t(1;19) without detectable E2A rearrangements in two t(14;18)-positive lymphoma/leukemia cases. Genes Chromosom Cancer 1994; 10:171-176.
23. Rowe D, Devaraj PE, Irving JAE et al. A case of mature B-cell ALL with coexistence of t(1;19) and t(14;18) and expression of the E2A/PBX1 fusion gene. Br J Haematol 1996; 94:133-135.
24. Berger R, Le Coniat M, Vecchione D et al. Cytogenetic studies of 44 T-cell acute lymphoblastic leukemias. Cancer Genet Cytogenet 1990; 44:69-75.

25. Kaneko Y, Frizzera G, Shikano T et al. Chromosomal and immunophenotypic patterns in T cell acute lymphoblastic leukemia (T ALL) and lymphoblastic lymphoma (LBL). Leukemia 1989; 3:886-892.
26. Secker-Walker LM, Campana D, Hawkins JM et al. Karyotype and T-cell receptor expression in T-lineage acute lymphoblastic leukemia. Genes Chromosom Cancer 1992; 4:41-45.
27. Raimondi SC, Behm FG, Roberson PK et al. Cytogenetics of childhood T-cell leukemia. Blood 1988; 72:1560-1566.
28. UKCCG. United Kingdom Cancer Cytogenetics Group. Translocations involving 9p and/or 12p in acute lymphoblastic leukemia. Genes Chromosom Cancer 1992; 5:255-259.
29. Pui C-H, Behm FG, Singh B et al. Heterogeneity of presenting features and their relation to treatment outcome in 120 children with T-cell acute lymphoblastic leukemia. Blood 1990; 75:174-179.
30. Chan LC, Kwong YL, Liu HW et al. Cytogenetic analysis of hematologic malignancies in Hong Kong. A study of 98 cases. Cancer Genet Cytogenet 1992; 62:154-159.
31. Carroll AJ, Crist WM, Link MP et al. The t(1;14)(p34;q11) is nonrandom and restricted to T-cell acute lymhoblastic leukemia: A Pediatric Oncology Group study. Blood 1990; 76:1220-1224.
32. Bash RO, Crist WM, Shuster JJ et al. Clinical features and outcome of T-cel acute lymphoblastic leukemia in childhood with respect to alterations at the TAL1 locus: A Pediatric Oncology Group study. Blood 1993; 81:2110-2117.
33. Brown L, Cheng JT, Chen Q et al. Site-specific recombination of the tal-1 gene is a common occurrence in human T cell leukemia. EMBO Journal 1990; 9:3343-3351.
34. Bash RO, Hall S, Timmons CF et al. Does activation of the TAL1 gene occur in a majority of patients with T-cell acute lymphoblastic leukemia? A pediatric oncology group study. Blood 1995; 86:666-676.
35. Breit TM, Mol EJ, Wolvers-Tettero IL et al. Site-specific deletions involving the tal-1 and sil genes are restricted to cells of the T cell receptor alpha/beta lineage: T cell receptor delta gene deletion mechanism affects multiple genes. J Exp Med 1993; 177:965-977.
36. Janssen JW, Ludwig WD, Sterry W et al. SIL-TAL1 deletion in T-cell acute lymphoblastic leukemia. Leukemia 1993; 7:1204-1210.
37. Robb L, Rasko JE, Bath ML et al. scl, a gene frequently activated in human T cell leukaemia, does not induce lymphomas in transgenic mice. Oncogene 1995; 10:205-209.
38. Breit TM, Beishuizen A, Ludwig WD et al. tal-1 deletions in T-cell acute lymphoblastic leukemia as PCR target for detection of minimal residual disease. Leukemia 1993; 7:2004-2011.

39. Jonsson OG, Kitchens RL, Baer RJ et al. Rearrangements of the tal-1 locus as clonal markers for T cell acute lymphoblastic leukemia. Journal of Clinical Investigation 1991; 87:2029-2035.
40. Aplan PD, Raimondi SC, Kirsch IR. Disruption of the SCL gene by a t(1;3) translocation in a patient with T cell acute lymphoblastic leukemia. J Exp Med 1992; 176:1303-1310.
41. Fitzgerald TJ, Neale GA, Raimondi SC et al. c-tal, a helix-loop-helix protein, is juxtaposed to the T-cell receptor-beta chain gene by a reciprocal chromosomal translocation: t(1;7)(p32;q35). Blood 1991; 78:2686-2695.
42. Whitlock JA, Raimondi SC, Harbott J et al. t(5;14)(q33-34;q11), a new recurring cytogenetic abnormality in childhood acute leukemia. Leukemia 1994; 8:1539-1543.
43. Lange BJ, Raimondi SC, Heerema N et al. Pediatric leukemia/lymphoma with t(8;14)(q24;q11). Leukemia 1992; 6:613-618.
44. Finger LR, Harvey RC, Moore RC et al. A common mechanism of chromosomal translocation in T- and B-cell neoplasia. Science 1986; 234:982-985.
45. Berkowicz M, Toren A, Rosner E et al. Lymphomatous T-cell leukemia in two Arab children. Is there a role for an environmental effect. Leukemia 1994; 8:1995-1998.
46. Dube ID, Raimondi SC, Pi D et al. A new translocation, t(10;14)(q24;q11), in T cell neoplasia. Blood 1986;67:1181-1184.
47. Park JK, Le Beau MM, Shows TB et al. A complex genetic rearrangement in a t(10;14)(q24;q11) associated with T-cell acute lymphoblastic leukemia. Genes Chromosom Cancer 1992; 4:32-40.
48. Lu M, Gong ZY, Shen WF et al. The tcl-3 proto-oncogene altered by chromosomal translocation in T-cell leukemia codes for a homeobox protein. EMBO Journal 1991; 10:2905-2910.
49. Hatano M, Roberts CW, Minden M et al. Deregulation of a homeobox gene, HOX11, by the t(10;14) in T cell leukemia. Science 1991; 253:79-82.
50. Dear TN, Sanchez-Garcia I, Rabbitts TH. The HOX11 gene encodes a DNA-binding nuclear transcription factor belonging to a distinct family of homeobox genes. Proceedings of the National Academy of Sciences of the United States of America 1993; 90:4431-4435.
51. Hawley RG, Fong AZ, Lu M et al. The HOX11 homeobox-containing gene of human leukemia immortalizes murine hematopoietic precursors. Oncogene 1994; 9:1-12.
52. Ribeiro RC, Raimondi SC, Behm FG et al. Clinical and biologic features of childhood T-cell leukemia with the t(11;14). Blood 1991; 78:466-470.

CHAPTER 7

Minor Translocation Subgroups, Partial Arm Deletions and Nonestablished Structural Change

TRISOMY OF CHROMOSOME 1q

There are three different types of rearrangement of the long arms of chromosome 1 which result in partial or total trisomy 1q. Duplication of part of the long arm of the chromosome—dup(1)(q12/23-q31/32)—results in partial trisomy. There are a number of derivative chromosomes involving unbalanced translocations of 1q, the most frequent being the unbalanced form of the translocation t(1;19),der(19)t(1;19). Finally a rare phenomenon, which has been termed "jumping translocations," involves an unbalanced translocation of 1q to a variety of different locations in several cell lines in the same patient.

Partial Duplication of 1q

Partial duplication of the long arms of chromosome 1—dup(1)(q12-21q31-32)—is found in association with two major subgroups in ALL, i.e., Burkitt's-type (BL) ALL and high hyperdiploidy. As secondary change to the Burkitt-type translocations in mature B ALL, partial duplication is found in 12% of cases. It has been found either in the stemline cell line (8% of cases) or in a sub-clone as evolutionary change (4% of cases).[1] Burkitt-type ALL has, until recently, had a uniformly poor prognosis. The association of dup(1)(q) with BL ALL and the observation that it may be found particularly in relapse may have been responsible for the reports of the particularly poor prognosis associated with dup(1)(q).[2-7] Confusingly, dup(1)(q) also occurs in high hyperdiploid clones with >50 chromosomes, i.e., the subgroup associated with the best prognosis.[8] In the LRF/UKCCG karyotype database (which excludes mature B ALL) dup(1)(q) has been seen in 47/1342 (3.5%) patients. None has been seen in the 24 infants and only one patient with dup(1)(q) was an adult: a 16 year old with a high hyperdiploid clone. The incidence in children was 46/1186 (3.9%) successfully karyotyped cases. Thirty-five patients had high hyperdiploid clones and three had 47-50 chromosomes (UKCCG database, see Appendix 2).

Trisomy of 1q by Unbalanced Translocations

This rearrangement is found particularly in association with t(1;19). More than half the cases in this subgroup have der(19)t(1;19) with two normal chromosomes 1. In addition, a significant number of patients (16%) have additional unbalanced translocations with further copies of 1q der(n)t(1;n)(q;n).[9]

Multiple Copies of 1q

A rare but recurring phenomenon seen in ALL is the presence of two or more clones each having derived chromosomes involving the whole of the long arm of 1. In general, breakage occurs in the heterochromatic region and subsequent joining is to the telomeric sequences of several different chromosomes. The resulting clones or cell lines have between them multiple copies of 1q. These have been termed "jumping translocations," a not altogether satisfactory name because it suggests that the same chromosome 1q attaches and then detaches itself at one site and then at another. A more likely mechanism is the repeated breakage of one homolog in a number of different cells, followed by translocation to different partner sites and the duplication of the remaining homolog. Each of these would be an independent event and the resulting progeny of each would form an independent clone. This instability of the heterochromatin on chromosome 1 may be related to exposure to viruses or may be a rare constitutional condition.[10-14]

TRANSLOCATION t(5;14)(q31;q32)

Translocation t(5;14) characterizes a subgroup of ALL with B-lineage immunophenotype and eosinophilia in the peripheral blood. It probably occurs in <1% of ALL patients. Grimaldi and Meaker cloned this translocation and have shown in two cases that it resulted in joining the promoter of *IL3* to the *IGH* gene, presumably increasing transcription of a structurally normal *IL3* gene. The authors subsequently observed excess *IL3* mRNA in the leukemic cells. The known effect of *IL3* on eosinophil differentiation might have contributed to the eosinophilia seen in these patients.[15,16]

ABNORMALITIES OF 6q

Abnormalities of the long arm of chromosome 6 (abn6q) have been found in between 6% and 11% of childhood ALL. The incidence in B lineage ALL is approximatley 8% which is significantly less frequent than the 12% described in T ALL ($p < 0.02$)[7,18-20] (see also Table 6.4). Secker-Walker et al, using a hierarchical classification for 350 adults, identified 4% of patients as del(6)(q).[22] Partial arm deletions account for more than 80% of these cases. Loss of sequences result from proximal breakpoints between 6q13 and q21 and distal breakpoints at q21 and q23 with common loss of band q21.[17,18] They rarely occur as the only change.

GENETICS OF del(6)(q)

The genetic consequences of abnormality have been examined in T ALL and are discussed in chapter 6.

PROGNOSIS

The prognostic significance of del(6)(q) in childhood has been variously shown to have a relatively good prognosis[21] or to have no impact on prognosis.[18] The true impact of this abnormality on prognosis is difficult to assess, since it frequently occurs with other chromosomal change. Secker-Walker et al show that the prognosis for adults with del(6)(q) was poor with 27% being estimated event-free survival at 3 years compared with an overall event-free survival of 36%.[22]

DELETIONS AND TRANSLOCATIONS OF 9p

In childhood, abnormalities of 9p (abn9p) occur with an incidence of 8%. Deletion accounts for 65% of these, the remainder occur as balanced or unbalanced translocations. Abnormalities of sequences from 9p12-p22 include loss of sequences by deletion or unbalanced translocation.[23-26] In all childhood series abn9p has it has been found more frequently in T than in C-ALL with a mean incidence of 13% and 8% respectively (p < 0.0.1) (see Table 6.5).[7,20,24,26] Secker-Walker et al classified 5% of 350 adults as abn9p. This was the only change in 4/350 (1%) of cases. Six patients (2%) had i(9)(q10) which results in loss of 9p and in two this was the only change. Translocation of 9p occurred to a number of different sites namely 3p21, 17p12, 5q13 and 13q32. One dic(9;12) was seen giving an incidence of <0.3% in this adult series.[22]

Genetic Involvement

The genetics of 9p deletions has been described in chapter 6.

Prognosis

In childhood, abnormalities of 9p appear to be without prognostic value, but failure involving CNS disease has been associated with this abnormality.[24,26] Adults classified as abn9p have been shown to have an intermediate prognosis with 48% event-free survival at three years.[22]

DELETIONS AND TRANSLOCATIONS OF 12p

Abnormalities of 12p in ALL were first highlighted by Williams et al in 1984.[27] In a later childhood series, an abnormality of 12p was found in approximately 10% of newly diagnosed cases.[28,29] Figure 7.1 shows a karyogram with del(12)(p12). Among 23 reported cases with 12p12 abnormalities, 18 had C-ALL, 3 had pre-B and 2 T ALL. No case had hyperdiploidy >50 and the median leukocyte count (30 x 10^9/l) was higher than in cases lacking these abnormalities.[28]

Abnormalities of 12p have included deletions, balanced and unbalanced translocations.[25] It is now clear that 12p may associate with a number of other sites in the genome to form reciprocal translocations. Among these several recurring translocations, includ-

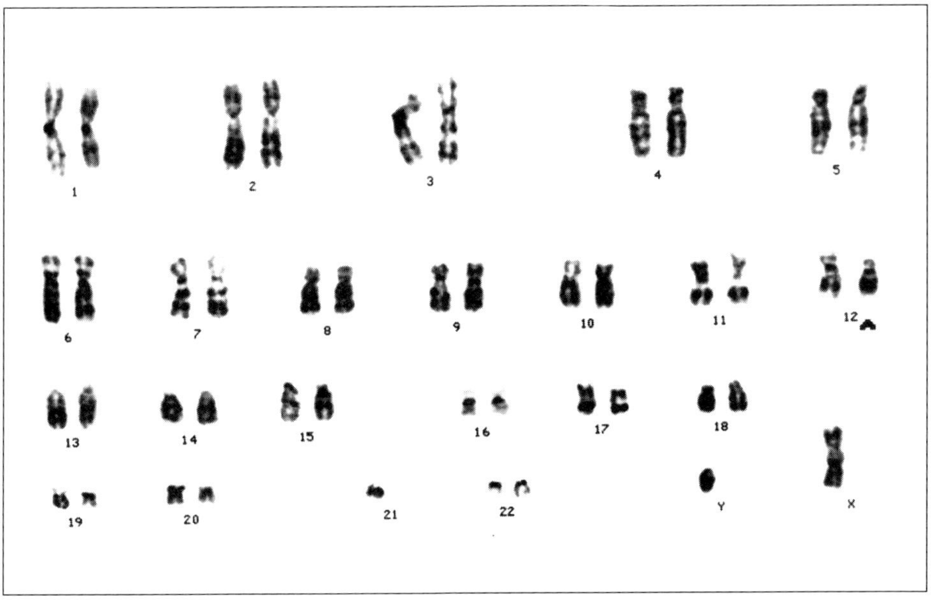

Fig. 7.1. Karyogram with 46, XY, del(12)(p12).

ing several dicentrics have been described. Partner sites to 12p are 7p11, 9p11-13 (usually as a dicentric), 13q14, 17q21 and 17p11-12. The t(12;17)(p13;q21) occurs in childhood early B lineage ALL with an incidence of <0.2%.[30]

Staining with trypsin-giemsa results in a pale band at the distal region of 12p making translocations of this region difficult to detect. It now seems likely that several cases described as del(12)(p) may turn out to be translocations. Romana et al investigated eight patients, thought to have a 12p deletion, by chromosome painting and showed three of them to have a t(12;21)(p11-12;q22) which had not been previously described.[31]

Genetic Consequences

Romana et al have demonstrated in two cases that t(12;21) resulted in the fusion of the genes *AML1* (located at 21q22) and the recently described *ETS* family gene, *TEL*, located at 12p13. Moreover the 12p breakpoints in both cases were localized within the same intron of the *TEL* gene. Both derivative chromosomes were transcriptionally active. The fused 5' *TEL-AML1* 3' was

it appears that stable dicentric chromosomes result from the presence of a damaged or inactive kinetochore. A similar mechanism may account for the nondisjunction of chromosomes which leads to trisomic clones.[45-48] The formation of stable dicentric chromosomes has been shown to be enhanced by radiation treatment but not by chemicals.[49]

dic(9;12)(p11-13;p11-12)

This translocation (seen in Fig. 7.3) identifies a subgroup of B-precursor ALL with an incidence of 8/432 in childhood.[50] A review and update of some 15 published cases with dic(9;12) showed an age range of 3-24 years (median seven years), WBC 0.8-132 median 8.2 x 10^9/L and an excess of males:females (12:3). These patients appeared to have an excellent prognosis with all 15 patients in continuous remission at 0-93 months with a median followup of 57+ months.[51,52] The formation of a dicentric chromosome as the only change results in a clone with 45 chromosomes. This translocation therefore identifies a good-risk group of patients with hypodiploidy.

ISOCHROMOSOMES

Isochromosomes have two arms which appear as mirror images of each other. They are formed by breakage at the centromere and the resulting chromosome retains only the long or short arms. Isodicentric chromosomes are also seen and it is presumed that these arise following breakage in the long or short arm and subsequent duplication of the whole structure.[53] There is at present no proof for the supposed mechanisms of isodicentric chromosome formation. There are at least four recurring isochromosomes in ALL: i(7)(q10), i(9)(q10), i(17)(q10) and more recently i(21)(q10) has been described as a new non random finding by Martineau et al.[54] In that study isochromosomes were found in 50/1053 (4.8%) patients with ALL. The incidence was twice as high in adults (9/116) (8%) as in children 41/919 (4%). Each isochromosome accounted for <1% of patients except i(21q10) which was found in 1.4% of patients. The incidence of isochromosomes in children reported by Pui et al was 83/2805 (3%) which was similar to the incidence reported by Martineau et al.[55]

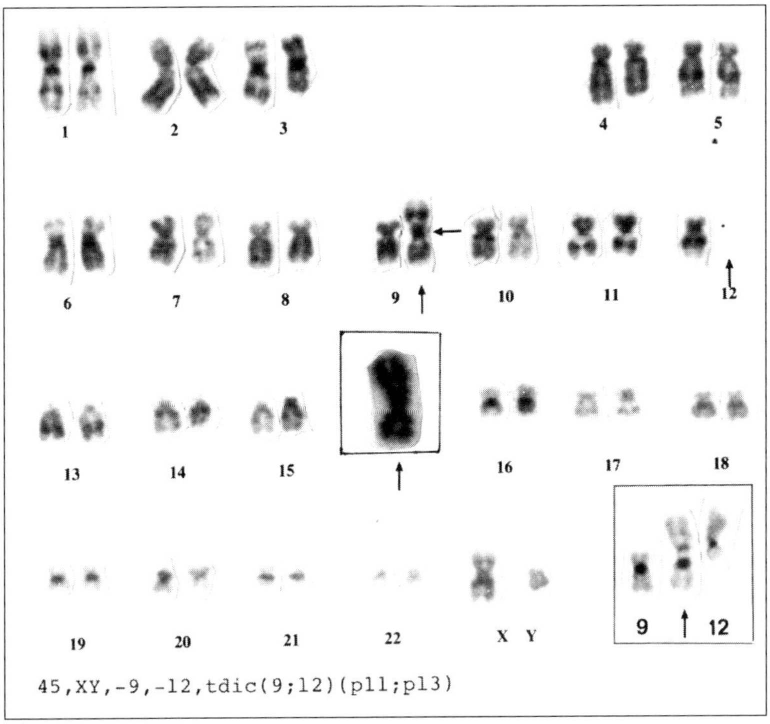

Fig. 7.3. Karyogram with 46, XY, dic(9;12)(p11;p11). Inserts show the translocation with two centromes G-banded (center) and C banded (bottom right) (loss of 12 was random).

Isochromosome 9q

Isochromosome 9q, i(9)(q10) is shown in Figure 7.4 and was recognized as a recurring abnormality in ALL in 1989 by Shippey et al.[56] Cases reported range from 3-35 years and WBC from 1.4->250 x 10^9/L. It has been seen in cases with blast cell immunophenotypes C-, Pre-B or T ALL. It is probably more common among childhood than adult cases The prognostic implications remain to be determined.[24]

Isochromosome 17q

This abnormality was reported in 8/469 (1.7%) of children with ALL. In all but one case this was found in a high hyperdiploid clone. Four of these patients relapsed early and it was suggested that the presence of i(17)(q10) represents a poor-risk finding in a hyperdiploid clone which is normally associated with a good prognosis.[57] More recently it has been shown that an excess of all

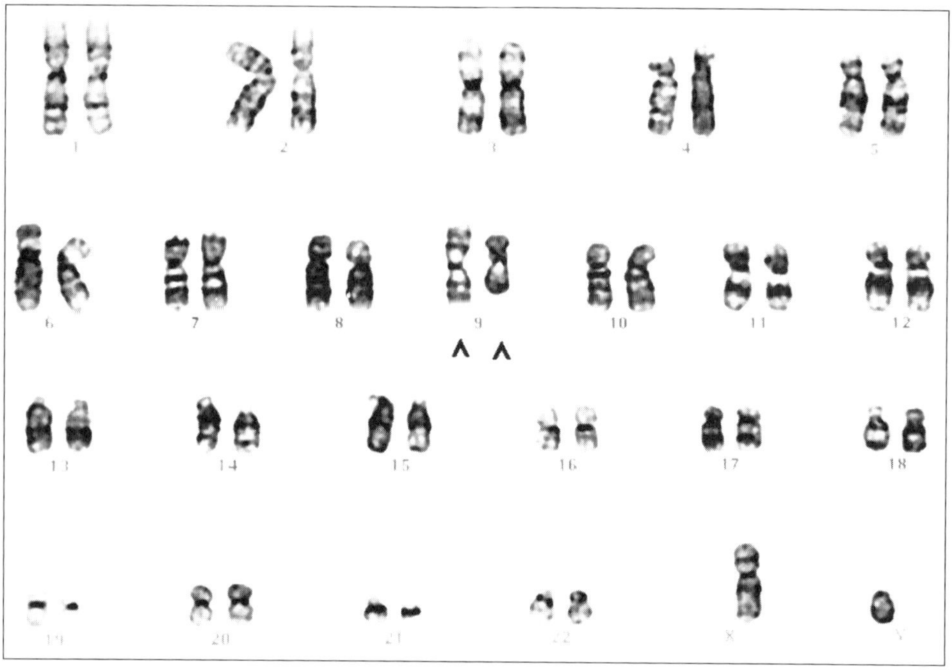

Fig. 7.4. Karyogram with 46, XY, del(9)(q22),i(9)(q10).

isochromosomes are found in hyperdiploid clones.[55] The study by Martineau et al concluded that this association was confined to isochromosomes 7 and 17.[54]

Isochromosome 21q

Martineau et al found isochromosome 21 at diagnosis in 15 patients aged 1-17 years. The median WBC was low (7 x 10^9/L). Chromosome 21 was the only chromosome involved in five cases. Some appeared as iso dicentrics, others as isochromosomes and others as both in the same patient. This group appeared to have a good prognosis compared with that of patients with other isochromosomes (projected disease-free survival at four years was 80%).[54]

NONESTABLISHED STRUCTURAL CHANGE

Apart from the chromosomal subgroups described in these chapters, structural chromosomal abnormalities found in ALL are remarkable for their heterogeneity. Among chromosomally abnormal cases, 82% of children and 94% of adults have structural change only 60% of which in each age group has been recognized as recurring. The remaining 40% of patients with structural change have chromosomally abnormal clones which appears to be a "one off" abnormality unique to that case.[58] Since a number of chromosomal subgroups with a very low incidence have been described, among the apparently unique abnormalities new chromosomal subgroups will undoubtedly emerge. Examples of one-off, clonal, chromosomal change are also referred to as "random" chromosomal change. It is however important to distinguish patients with these clonal abnormalities from patients with random abnormal cells which could not be identified as clonal. This considerable heterogeneity found in malignancy shows that there are many different cytogenetic configurations which confer a proliferative advantage on the cell or result in a failure of apoptosis (programmed cell death). It therefore appears that genes with oncogenic potential are abundantly present throughout the genome and the creation of aberrant fusion proteins may result from chromosomal rearrangements of many different locations. The surprise is not that this process happens at all but rather that it does not happen more frequently.

Translocations and Prognosis

The relevance of translocations to prognosis is now uncertain. Initially it was shown that translocations were an adverse prognostic feature.[21,59] Others have concluded that this is not the case.[7,37,60] Indeed Rubin et al have shown that among 146 children with ALL there was no difference in survival between patients with and without translocations (73% at four years for both groups)[60] provided that patients with an established poor-risk translocation were removed.[7] It has for many years been appreciated that certain translocations have a particularly poor prognosis and the presence of these in any "translocation" group could mask any neutral or even beneficial effect of smaller subgroups. Studies from St Jude's Children's Hospital originally showed that the presence of structural change in a high hyperdiploid clone compromised the otherwise excellent prognosis of this subgroup.[8] The same observation was made by Van der Plas.[20] The most recent data from St Jude's suggest that this difference is no longer apparent.[61] Translocations are many and diverse. An important consideration for the future is not whether any translocation is present but whether the particular translocation in question is associated with a good or poor response to treatment.

SUMMARY

The chromosomal abnormalities described in this chapter include abnormalities which are not associated with well-defined clinical features or a particular prognosis. These include partial duplication of the long arm of chromosome 1, involvement of chromosome arms 6q, 9p and 12p as either deletions or translocations, dicentric chromosomes and isochromosomes.

REFERENCES

1. Mitelman F. Catalog of chromosome aberrations in cancer. 5th ed. New York:Wiley-Liss, 1994.
2. Berger R, Bernheim A. Cytogenetic studies on Burkitt's lymphoma-leukemia. Cancer Genet Cytogenet 1982; 7:231-244.
3. Berger R, Bernheim A. Cytogenetics of Burkitt's lymphoma-leukaemia: a review. IARC Sci Publ 1985; 65-80.
4. Slater RM, Behrendt H, van Heerde P. Cytogenetic studies on four cases of non-endemic Burkitt lymphoma. Med Pediatr Oncol 1982; 10:71-84.
5. Knuutila S, Elonen E, Heinonen K et al. Chromosome abnormalities in 16 Finnish patients with Burkitt's lymphoma or L3 acute lymphocytic leukemia. Cancer Genet Cytogenet 1984; 13:139-151.
6. Secker-Walker LM, Stewart E, Norton J et al. Multiple chromosome abnormalities in a drug resistant TdT positive B-cell leukemia. Leukemia Research 1987; 11:155-161.
7. GFCH. Groupe Francais de Cytogenetique Hematologique: Collaborative study of karyotypes in childhood acute lymphoblastic leukemias. Leukemia 1993; 7:10-19.
8. Pui CH, Raimondi SC, Dodge RK et al. Prognostic importance of structural chromosomal abnormalities in children with hyperdiploid (greater than 50 chromosomes) acute lymphoblastic leukemia. Blood 1989; 73:1963-1967.
9. Secker-Walker LM, Berger R, Fenaux P et al. Prognostic significance of the balanced t(1;19) and unbalanced der(19)t(1;19) translocations in acute lymphoblastic leukemia. Leukemia 1992; 6:363-369.
10. Raimondi SC, Ragsdale ST, Behm F et al. Multiple telomeric associations of a trisomic whole q arm of chromosome 1 in a child with acute lymphoblastic leukemia. Cancer Genet Cytogenet 1987; 24:87-93.

11. Shippey CA, Layton M, Secker-Walker LM. Leukemia characterized by multiple sub-clones with unbalanced translocations involving different telomeric segments: Case report and review of the literature. Genes Chromosom Cancer 1990; 2:14-17.
12. Whang Peng J, Lee EC, Sieverts H et al. Burkitt's lymphoma in AIDS: cytogenetic study. Blood 1984; 63:818-822.
13. Fitzgerald PH, Morris CM. Telomeric association of chromosomes in B-cell lymphoid leukemia. Hum Genet 1984; 67:385-390.
14. Morris CM, Fitzgerald PH, Neville MA et al. Does multisomy of chromosome 1q confer a proliferative advantage in B-cell acute lymphoblastic leukemia? Cancer 1984; 54:48-53.
15. Grimaldi JC, Meeker TC. The t(5;14) chromosomal translocation in a case of acute lymphocytic leukemia joins the interleukin-3 gene to the immunoglobulin heavy chain gene. Blood 1989; 73:2081-2085.
16. Meeker TC, Hardy D, Willman C et al. Activation of the interleukin-3 gene by chromosome translocation in acute lymphocytic leukemia with eosinophilia. Blood 1990; 76:285-289.
17. Barletta C, Pelicci PG, Kenyon LC et al. Relationship between the c-myb locus and the 6q-chromosomal aberration in leukemias and lymphomas. Science 1987; 235:1064-1067.
18. Hayashi Y, Raimondi SC, Look AT et al. Abnormalities of the long arm of chromosome 6 in childhood acute lymphoblastic leukemia. Blood 1990; 76:1626-1630.
19. Raimondi SC, Behm FG, Roberson PK et al. Cytogenetics of childhood T-cell leukemia. Blood 1988; 72:1560-1566.
20. Van der Plas DC, Hahlen K, Hagemeijer A. Prognostic significance of karyotype at diagnosis in childhood acute lymphoblastic leukemia. Leukemia 1992; 6:176-184.
21. IWCL3 . Third International Workshop on Chromosomes in Leukemia (1980): Clinical significance of chromosomal abnormalities in acute lymphoblastic leukemia. Cancer Genet Cytogenet 1981; 4:111-137.
22. Secker-Walker LM, Prentice HG, Durrant J et al. Cytogenetics adds independent prognostic information in adults with acute lymphoblastic leukaemia on MRC trial UKALL XA. Br J Haematol 1997; in press.
23. Carroll AJ, Castleberry RP, Crist WM. Lack of association between abnormalities of the chromosome 9 short arm and either "lymphomatous" features or T cell phenotype in childhood acute lymphocytic leukemia. Blood 1987; 69:735-738.
24. Pollak C, Hagemeijer A. Abnormalities of the short arm of chromosome 9 with partial loss of material in hematological disorders. Leukemia 1987; 1:541-548.

53. De la Chapelle A. How do human isochromosomes arise? Cancer Genet Cytogenet 1982; 5:173-179.
54. Martineau M, Clark R, Farrell DM et al. Isochromosomes in Acute Lymphoblastic Leukaemia: i(21q) is a Significant Finding. Genes Chromosom Cancer 1996; 17:21-30.
55. Pui C-H, Carroll AJ, Raimondi SC et al. Isochromosomes in childhood acute lymphoblastic leukemia: a collaborative study of 83 cases. Blood 1992; 79:2384-2391.
56. Shippey CA, Lawlor E, Secker-Walker LM. Isochromosome 9q in acute lymphoblastic leukemia: a new non-random finding. Leukemia 1989; 3:195-199.
57. Pui CH, Raimondi SC, Williams DL. Isochromosome 17q in childhood acute lymphoblastic leukemia: an adverse cytogenetic feature in association with hyperdiploidy? Leukemia 1988; 2:222-225.
58. Pui CH, Crist WM, Look AT. Biology and clinical significance of cytogenetic abnormalities in childhood acute lymphoblastic leukemia. Blood 1990; 76:1449-1463.
59. Williams DL, Harber J, Murphy SB et al. Chromosomal translocations play a unique role in influencing prognosis in childhood acute lymphoblastic leukemia. Blood 1986; 68:205-212.
60. Rubin CM, Le Beau MM, Mick R et al. Impact of chromosomal translocations on prognosis in childhood acute lymphoblastic leukemia. Journal of Clinical Oncology 1991; 9:2183-2192.
61. Raimondi SC, Pui C-H, Hancock ML et al. Heterogeneity of hyperdiploid (51-67) childhood acute lymphoblastic leukemia. Leukemia 1996; 10:213-224.

CHAPTER 8

Cytogenetics of Infants and Adolescents with ALL

ACUTE LEUKEMIA IN INFANTS

Patients diagnosed at or under the age of one year are classified as infants. Acute leukemia is rare in this age group. It may be of the myeloid or lymphoid variety and frequently shows evidence of stem cell involvement in displaying features of both types of cell. The blasts in these cases are either biphenotypic, expressing both lymphoid and myeloid antigens on the same cell, or bilineage, in which antigens of the myeloid and lymphoid lineages are expressed on different cells. In practice, the distinction between biphenotypic and bilineage may not always be clear. The cytogenetic abnormalities found in infants with acute leukemia, whether lymphoid or myeloid, are characterized by a high incidence of pseudodiploidy and, in particular, structural change which is likely to be complex and to include a structural abnormality of 11q23.[1,2] Infants with ALL tend to have hepatosplenomegaly, leukocytosis, frequent nervous system involvement and most succumb to the disease.

CYTOGENETICS IN INFANTS

The incidence of different cytogenetic subgroups in ALL taken from six publications is shown in Table 8.1.[1,3-7] From this it can be seen that the most common finding is an abnormality of 11q23 which has been found in over 60% of cases. Other chromosomal abnormalities or chromosomally normal cases each account for approximately 20% of cases. Table 8.2 shows that infant ALL

Table 8.1. Incidence of different cytogenetic abnormalities in 224 infants with ALL

Total	Normal	t(4;11)	t(9;11)	t(11;19)	Other 11q	Other	Refs
17	1	6	2	7	1	0	Katz et al[3]
15	6	6				3	Ferster et al[6]
22	5	13	0	1	0	3	Chen et al[5]
39	11	12	2	1	2	11	Heerema et al[1]
27	2	6	3	4	4	8	Pui et al[4]
78	13	37	1	7	6	14	Rubnitz et al[17]
26	3	11	2	0	3	7	Appendix 2
224	41	91	10	20	16	46	
%	18.3	40.6	4.5	8.9	7.1	20.4	

occurs more frequently in females than in males and in addition that females are more likely than males to have an abnormality at 11q23 ($p < 0.05$).[1,3-5]

INFANTS WITH t(4;11)(q21;q23)

The most frequent chromosomal abnormality in infant ALL is t(4;11)(q21;q23) (see chapter 4). It accounts for 40% of published infant cases and as in other age groups is associated with female sex, high white blood cell counts (>100 x 10^9 /L) and a pre-pre-B or null cell (CD19+ve,CD10-ve) immunophenotype. Four recent publications indicate that the male:female ratio in infants is 17:20 (see Table 8.2). Heerema et al showed an excess of infants with t(4;11) to be under rather than over the age of six months (10:2),[1] but in two other publications the reverse ratio was seen (4:8) (see Table 8.3).[3,4]

Table 8.2. Distribution of sex by chromosomal subgroup of 132 infants with ALL

Total Cases	Normal M:F	t(4;11) M:F	t(9;11) M:F	t(11;19) M:F	Other 11q M:F	Other M:F	Refs
17	1:0	3:3	1:1	2:5	1:0	0	Katz et al[3]
22	4:1	5:8	0	0:1	0	2:1	Chen et al[5]
39	7:4	5:7	0:2	1:0	0:2	3:8	Heerema et al[1]
28	1:1	4:2	0:3	1:4	1:3	5:3	Pui et al[4]
26	1:2	4:7	0:2	0	0:3	6:1	Appendix 2
	14:8	21:27	1:8	4:10	2:8	16:13	Total

Molecular investigations of *MLL/AF4* in infants

Borkhardt et al[8] detected the molecular fusion brought about by t(4;11) *MLL/AF4*, by RT-PCR in 5/10 infants with ALL. Various splicing variants were demonstrated. Three patients had fusion between *MLL* exon 6 and codon 362 of the known *AF4* cDNA sequence. In two patients, more than one specific RT-PCR product was obtained, possibly due to alternative splicing. In one case, *MLL* exon 7 or 8 was fused to *AF4* codon 348. In the other case with alternative splicing, *MLL* exon 7 or 8 was fused to *AF4* but this time to codon 362. None of the 10 cases investigated showed the reciprocal mRNA transcript *AF4/MLL*.

t(9;11)(p21;q23) AND t(11;19)(q23;p13)

Two common translocations of 11q23 in infant ALL are t(9;11)(p21;q23) found in 5% of cases and t(11;19) found in 9% of cases (see Table 8.1). The translocation t(11;19)(q23;p13) was first described in a four month old girl with CD19+ congenital ALL by Heim et al in 1986.[9] These and other translocations of 11q23 appear to be even more strongly associated with female sex than t(4;11). Male to female ratios of published cases have been: 1:6 for t(9;11), 4:10 for t(11;19) and 2:8 for other translocations of 11q23 (see Table 8.2). All these abnormalities of 11q23 have

Table 8.3. Distribution by age of chromosomal subgroups in 70 infants with ALL

Total	Normal <6m:>6m	t(4;11) <6m:>6m	t(9;11) <6m:>6m	t(11;19) <6m:>6m	Other 11q <6m:>6m	Other <6m:>6m	Refs
17	1:0	2:4	1:1	6:1	1:0	0	Katz et al[3]
27	0:2	2:4	2:1	4:0	2:2	2:6	Pui et al[4]
26	1:2	6:5	1:1	0:0	2:1	1:6	Appendix 2
70	2:4	10:13	4:3	10:1	5:3	3:12	Total

been found predominantly in infants under six months of age (see Table 8.3).[3,4]

Abnormalities of 11q23 found in pairs of twins have contributed to our understanding of the etiology of leukemia. Four pairs of monozygotic twins, each of whom developed ALL with t(11;19)(q23;p13.3) as infants, have been described. In each case, the *MLL* gene showed an identical nonconstitutional rearrangement. From this, the authors suggested that a transformation event which occurs in utero in one twin could be passed to the other. In one twin pair, the placenta was diplacental/dichorionic which indicated that the transfer of malignant cells could have occurred by crossing to the maternal circulation and back to the second twin.[10,11] Nonidentical *IGH* rearrangements found in one pair were taken to indicate that t(11;19) had arisen in an early B cell before the *IGH* gene was rearranged.[11]

Other Chromosomal Findings in Infant ALL

Approximately 20% of infants have chromosomal abnormalities other than of 11q23. Amongst the 25 published karyotypes with other abnormalities, a new nonrandom finding was identified in four cases among 103 infants, namely t(5;15)(p15;q11/12) (4%). Three infants had translocations of Xp (3%). Established change found in these infants included: high hyperdiploidy (3%), deletion of 6q (2%), and one each with der(19)t(1;19) and dic(9;12) (1%).[1,4-6] Infants with these other abnormalities were more likely to be male (sex ratio 16:13), and in contrast to the 11q23 cases were mostly over six months of age.[1,3,4] The 20% of infants who have been reported as chromosomally normal have been mostly male (20:10) and over six months of age (see Tables 8.2 and 8.3).[1,3,4]

Simple Karyotypes and Additional Change

Published karyotypes in three series of 103 infants with ALL[1,4,5,12] showed that among 40 infants with t(4;11) this was the only change in 24 patients and was accompanied by other changes either in the stem-line or in a side-line in the remaining 16 cases. The translocation t(9;11) was the only change in 1/4 (25%) cases, t(11;19) was the only change in 10/13 (77%) of cases and other abnormalities of 11q23 occurred as the only change in 4/7 (57%) cases. Thus, 11q23 abnormalities together are seen as the only change in 39/64 (61%) cases. Other clonal change in infant ALL showed a translocation or deletion as the only structural change in 6/15 (40%) patients and five cases showed numerical gain only. The distribution of 77 chromosomally abnormal infant karyotypes by ploidy, taken from the same publications, was hypodiploid: 3/77 (4%), pseudodiploid 64/77 (83%), low hyperdiploid 7/77 (9%) and high hyperdiploidy 3/77 (4%). The distribution was different for clones with and without 11q23 abnormalities. Those with an abnormal 11q23 were either pseudodiploid 51/55 (93%) or low hyperdiploid in 4/55 (7%). This contrasted with the remaining cases with an abnormal clone which had pseudodiploidy in 13/22 (58%) cases. Hypodiploidy, low hyperdiploidy and high hyperdiploidy, each accounted for 3/22 (14%) cases.

Prognosis

Several authors have commented on the poor prognosis of all infants but this has been particularly marked for patients with 11q23 abnormalities.[1,3,4,6] Pui et al[4] show the three year event-free survival for infants with and without 11q23 involvement to be 13% and 67%, respectively (see Fig. 8.1). Heerema et al[1] showed a poorer survival for t(4;11) patients than for those without (p = 0.009) (see Fig. 8.2).

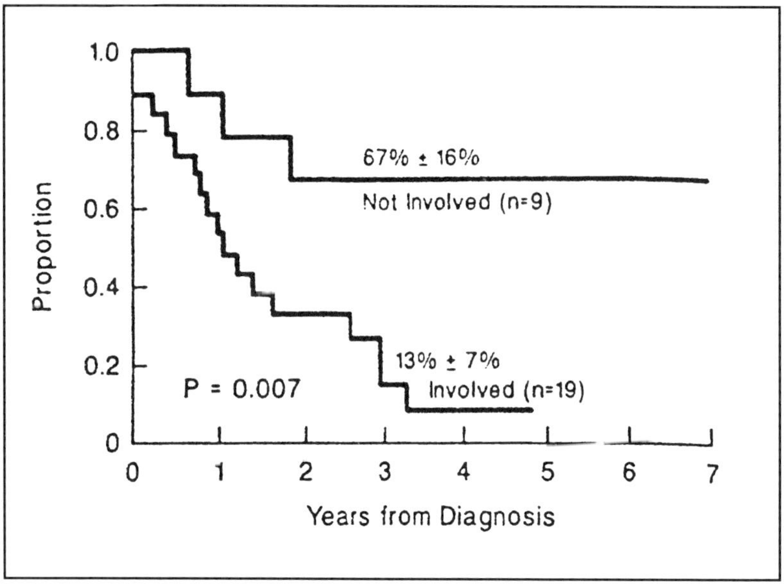

Fig. 8.1 Event-free survival for infants with (involved) or without (not involved) 11q23/MLL rearrangements. Note the poor outcome of infants with 11q23/MLL involvement. Reprinted from Pui et al.[4]

Incidence Of *Mll* Gene Rearrangement

Molecular investigations of infant ALL have consistently shown a higher incidence of rearrangement of the *MLL* gene than that of 11q23 involvement demonstrated cytogentically. Of particular interest were seven cases reported by Pui et al[4] in which rearrangement of the *MLL* gene was found in patients with karyotypic abnormalities other than 11q23 (five cases), in a chromosomally normal case and in one case in which cytogenetics had failed. Chen et al and Rubnitz et al have demonstrated rearrangement of the *MLL* (*HRX*) gene by Southern blotting in 21/30 (70%) and 78/96 (81%) infants, respectively.[5,7] In each study rearranged *MLL* was associated with a significantly higher median WBC (497 or 173 x 10^9/L v 33 or 52 x 10^9/L), with female sex and with age under six months. Event-free survival at three years for these patients was 15% or 19% compared with 80% or 46% survival for those with germ line *MLL*.

Biology of Infant ALL

Biological features of the blasts in infant ALL may help to explain their poor survival. Blasts survive in vitro on stroma, giving

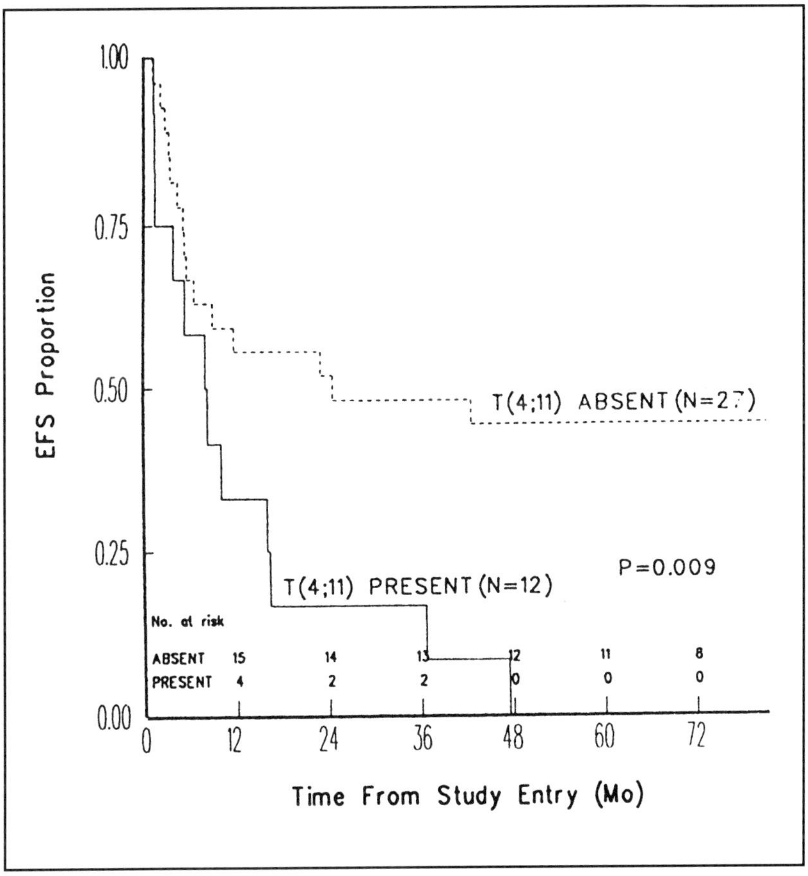

Fig. 8.2. Event-free survival of infants with and without t(4;11). Reprinted from Heerema et al.[1]

92% recovery after seven days. It has been shown in particular that t(4;11) blasts will replicate independently of growth factors.[13] An apparently contradictory feature of infant ALL is that cytotoxic drugs have been shown to clear from the patient more slowly than from older patients. The blasts will therefore have prolonged exposure to the drugs but nevertheless survive, suggesting that they are less sensitive to the drugs than are the blasts of older patients.

ADOLESCENTS

Adolescents (patients aged between 10 and 20 years) have also been shown to represent a poor-risk group within the general classification of childhood ALL. Patients aged 10-21 years when compared with patients aged 1-10 years have been shown to have more adverse prognostic features. These include T ALL, L2 blast cell morphology, higher leukocyte count and leukemic cell DNA index <1.16.[14]

Cytogenetics in Adolescents

A difference in the distribution of patients between the good- and poor-risk cytogenetic subgroups, high hyperdiploidy and Ph+ has also been noted. Thus hyperdiploidy >50 was found in 190/498 (38%) of childhood cases compared with 30/157 (19%) of adolescents, and Ph was seen in 14/383 (3.7%) children and in 9/127 (7%) of adolescents.[14]

SUMMARY

In summary, >60% of infants with ALL have an abnormality of 11q23, involving rearrangement of the *MLL* gene. Over half of these patients have t(4;11)(q21;q23) and the others have translocations between 11q23 and one of a number of other sites in the genome. Patients with involvement of 11q23 are more likely to be female and under 6 months old. Infants with other chromosomal findings at diagnosis are more likely to be male and to be aged >6 months. The cytogenetics of adolescents with ALL have been little studied. However, there is evidence that the incidences of good-risk, high hyperdiploid cases and poor-risk, Philadelphia positive patients are intermediate between those in childhood and adults.

References

1. Heerema NA, Arthur DC, Sather H et al. Cytogenetic features of infants less than 12 months of age at diagnosis of acute lymphoblastic leukemia: impact of the 11q23 breakpoint on outcome: a report of the Childrens Cancer Group. Blood 1994; 83:2274-2284.
2. Lampert F, Harbott J, Ritterbach J. Cytogenetic findings in acute leukaemias of infants. British Journal of Cancer—Supplement 1992; 18:S20-S22.
3. Katz F, Ball S, Gibbons B. Acute lymphoblastic leukaemia in infancy: clinical and biological features. Leukemia and Lymphoma 1990; 2:259-269.
4. Pui CH, Behm FG, Downing JR et al. 11q23/MLL rearrangement confers a poor prognosis in infants with acute lymphoblastic leukemia. Journal of Clinical Oncology 1994; 12:909-915.

5. Chen CS, Sorensen PH, Domer PH et al. Molecular rearrangements on chromosome 11q23 predominate in infant acute lymphoblastic leukemia and are associated with specific biologic variables and poor outcome. Blood 1993; 81:2386-2393.
6. Ferster A, Bertrand Y, Benoit Y et al. Improved survival for acute lymphoblastic leukaemia in infancy: the experience of EORTC-Childhood Leukaemia Cooperative Group. Br J Haematol 1994; 86:284-290.
7. Rubnitz JE, Link MP, Shuster JJ et al. Frequency and prognostic significance of HRX rearrangements in infant acute lymphoblastic leukemia: a Pediatric Oncology Group study. Blood 1994; 84:570-573.
8. Borkhardt A, Repp R, Haupt E, Brettreich S, Buchen U, Gossen R, Lampert F. Molecular analysis of MLL-1/AF4 recombination in infant acute lymphoblastic leukemia. Leukemia 1994; 8:549-553.
9. Heim S, Kristoffersson U, Mandahl N et al. Reciprocal translocation (11;19)(q23;p13) in congenital acute lymphoblastic leukemia. Cancer Genet Cytogenet 1986; 23:239-244.
10. Ford AM, Ridge SA, Cabrera ME et al. In utero rearrangements in the trithorax-related oncogene in infant leukaemias. Nature 1993; 363:358-360.
11. Gill Super HJ, Rothberg PG, Kobayashi H et al. Clonal, nonconstitutional rearrangements of the MLL gene in infant twins with acute lymphoblastic leukemia: in utero chromosome rearrangement of 11q23. Blood 1994; 83:641-644.
12. Gibbons B, Katz FE, Ganly P et al. Infant acute lymphoblastic leukaemia with t(11;19). Br J Haematol 1990; 74:264-269.
13. Manabe A, Coustan-Smith E, Behm FG et al. Bone marrow-derived stromal cells prevent apoptotic cell death in B-lineage acute lymphoblastic leukemia. Blood 1992; 79(9):2370-2377.
14. Santana VM, Dodge RK, Crist WM et al. Presenting features and treatment outcome of adolescents with acute lymphoblastic leukemia. Leukemia 1990; 4:87-90.

CHAPTER 9

The Use of Molecular Techniques for the Detection of Cytogenetic Abnormalities and Their Impact on Patient Management

When acquired chromosomal abnormalities were first described in hematological malignancies during the 1960s, it was unclear whether these were epiphenomena or initiating events in the etiology of neoplasia. The turning point in our understanding of the relationship between chromosomal abnormalities and leukemogenesis was an observation by Rowley in 1984.[1] Rowley observed that translocation breakpoints in leukemia frequently coincided with the chromosomal location of a cellular oncogene. Subsequent investigations have led to an unravelling of the genetic events brought about by chromosomal translocations. These result either in the downregulation of a cellular oncogene or in the creation of an entirely novel, leukemia-specific gene with a novel mRNA which is directly involved in leukemogenesis.

CHROMOSOMALLY-MEDIATED GENETIC CHANGE

Translocations

It is now well established that translocations contribute to leukemogenesis by bringing about aberrant activation of a gene. Most of these activated genes are transcription factors. There are two ways in which activation can occur. The first method is found only in leukemias of mature B cells (B-ALL) and in T ALL. It occurs when chromosomal breakage and rejoining results in the juxtaposition of one of the immunoglobulin (*IG*) or T cell receptor (*TCR*) genes with the oncogene c-*MYC* located at 8q24 or another putative oncogene. In T and B ALL, *MYC* comes under the influence of the active *IG* or *TCR* locus which stimulates transcription of the oncogene, leading to deregulation, increased transcription and neoplastic growth. In these cases, the breakpoints lie outside the oncogene which may be juxtaposed either upstream or down stream of the *IG* or *TCR* gene, suggesting that the promoters involved in transcription may work in either direction. A more widespread phenomenon is the activation of a putative oncogene by breakage within two normally disparate genes followed by inappropriate rejoining. This results in the fusion of exons to form a chimeric gene product in which the transcriptional sequences of one gene, not normally expressed in hematopoeitic tissue, come under the control of the promoter regions of another gene. This leads to genetic downregulation and the formation of an aberrant transcription factor. See Rabbitts for a review.[2]

Deletions, Duplications and Numerical Change

Deletions or chromosome loss and duplications or chromosome gain result in loss and gain, respectively, of some or all of the genes on a chromosome. Thus there will be an alteration in the number of copies of the genes located on the region involved.

INTERACTION OF CYTOGENETIC AND MOLECULAR TECHNIQUES

The development of molecular techniques during the past decade has resulted in a two way exchange between cytogeneticists and molecular biologists. Cytogenetic abnormalities help to locate regions of the genome involved in leukemogenesis; molecular techniques are needed to clone and sequence the genes. Their location in the genome can then be verified by a combination of cytogenetic and molecular methods. This chapter provides an overview of the impact of molecular techniques on cytogenetic investigations and their role in diagnosis and patient management.

Both structural and numerical chromosomal abnormalities are amenable to detection by molecular techniques, provided the appropriate probes are available. In acute lymphoblastic leukemia (ALL), probes are available for the detection of certain translocations, some partial arm deletions, any whole chromosome gain (trisomy) and any whole chromosome loss (monosomy). Furthermore, it is now possible to identify or clarify marker chromosomes and to assign breakpoints in chromosomal rearrangements to specific genes or regions of DNA.

There are three major techniques which enable the investigation of chromosomally-mediated genetic change in malignant cells: i.e., Southern blotting, reverse transcription polymerase chain reaction (RT-PCR) and fluorescence in situ hybridization (FISH).

SOUTHERN BLOTTING

Translocations between chromosomes result from breakage within two chromosomes in the same cell at the same time followed by the reattachment of each deleted fragment to the wrong chromosome. If breakage occurs within a gene, thereby physically disrupting, it this can be investigated by Southern blotting. Southern blotting is a process by which DNA is extracted from the malignant tissue, digested with site-specific restriction enzymes and the resulting fragments separated by gel electrophoresis. The DNA fragments are then transferred and attached to a nylon or nitrocellulose filter. A radio-labelled DNA probe, specific to the gene in question, is then hybridized to the filter. Autoradiography is then used to visualize the appropriate DNA fragments. If breakage has occurred within the gene, the gel will show two different sized fragments. One is the germline fragment located on the intact chromosome and the other is the cut or rearranged fragment of the gene on the chromosome which has contributed to the translocation. This technique has a sensitivity of between 5% and 20%.

REVERSE TRANSCRIPTION-POLYMERASE CHAIN REACTION (RT-PCR)

The polymerase chain reaction (PCR) is a way of producing large amounts of a specific DNA fragment of defined length and sequences from small amounts of source material. Reverse transcription-polymerase chain reaction (RT-PCR) is a way of obtaining cDNA from RNA and amplifying that by PCR. Translocations result from the breakage and subsequent rejoining of DNA sequences from two different chromosomes. If both breaks occur within genes and the rejoining results in gene fusion to produce a new chimeric gene with a novel mRNA and if the sequences adjacent to the breakpoints of both genes are known, these sequences can be used as primers to amplify the gene product(s). This technique will detect aberrant gene sequences present in minute quantities, enabling the detection of approximately 1 cell in 100,000.

FLUORESCENCE IN SITU HYBRIDIZATION (FISH)

There are now available whole chromosome probes, chromosome-specific alpha satellite centromeric probes and gene-specific probes. Also available are techniques for labelling sequences within the probes with fluorescent nucleotides. Thus it is now possible to investigate numerical chromosomal change as well as genetic rearrangements by fluorescence in situ hybridization (FISH).[3] FISH relies on the ability of fluorescent molecules to be deposited in chromatin at the sites of specific DNA sequences. DNA or RNA sequences are first labelled with reporter molecules. The probe and the target chromosomes or nuclei are denatured. Complementary sequences in the probe and target are then allowed to reanneal. After washing and incubation in fluorescently labelled affinity reagents, a discrete fluorescent signal is observed at the site of probe hybridization.

FISH, to a cytogeneticist, is the most versatile and most informative of the molecular techniques. DNA sequences can be localized both in metaphase spreads and in interphase nuclei. Sequences which can be highlighted range from the entire human genome, through entire chromosomes, 6 chromosomal subregions and single copy sequences. There are a variety of probe labelling schemes available for simultaneous detection of a number of sequences, which will allow detection in up to 24 different colors, enabling visualization of each of the 24 human chromosomes in a single cell. The availability of commercially prepared labelled probes, high quality fluorescent microscopes, filters, cameras and software packages makes FISH a reliable and user-friendly technique for analyzing malignant tissue. FISH can be used to detect extra, missing or broken and rearranged copies of chromosomes as well as to locate genes on metaphase chromosomes and detect them in interphase nuclei. The sensitivity of interphase FISH ranges from 5% for a simple monosomy to 10^{-4} for triple trisomy. Figure 9.1 shows the relative sensitivity of molecular and cytogenetic techniques.

DETECTION OF CLONES AT DIAGNOSIS WHEN CYTOGENETICS FAILS

Molecular techniques are a useful adjunct to cytogenetic investigations at diagnosis and are of particular benefit when cytogenetics is unavailable or has failed.

SOUTHERN BLOTTING AND RT-PCR

Southern blotting and RT-PCR have the advantage that they can be carried out on DNA or RNA prepared from the total cell population of the leukemic sample and do not require the preparation of cells at metaphase. Translocations which result in breakage within a well-defined region of one gene can be investigated by Southern blotting and translocations resulting in gene-fusion and the creation of an aberrant mRNA can be investigated by RT-PCR. The proviso here is that the relevant probes must be available. At the present time the cloned and sequenced translocations in ALL comprise t(1;19)(q21;p13) and the variant der(19)t(1;19), t(17;19)(q21;p13), t(4;11)(q21;q23), t(9;22)(q34;q11), t(12;21)(p13;q22) and some of the variant 11q23 translocations such as t(1;11), t(10;11) and t(11;19). These abnormalities can be investigated either by Southern blotting to detect the rearrangement of one or other of the genes involved or by RT-PCR to detect the fused gene sequences and aberrant mRNA. Southern blotting can be used if only one of the breakpoints has been cloned. For example abnormalities of 11q23, for which the partner gene has not yet been cloned, can be investigated for rearrangement of the *MLL* gene by Southern blotting (see Table 9.1).

FISH

Centromeric probes are ideally suited to the detection of whole chromosome gains as trisomies, tetrasomies or greater. False positive figures for trisomies are reasonably low ranging from 0.5% to 4% depending on the chromosome.[4,5] This is not a problem at

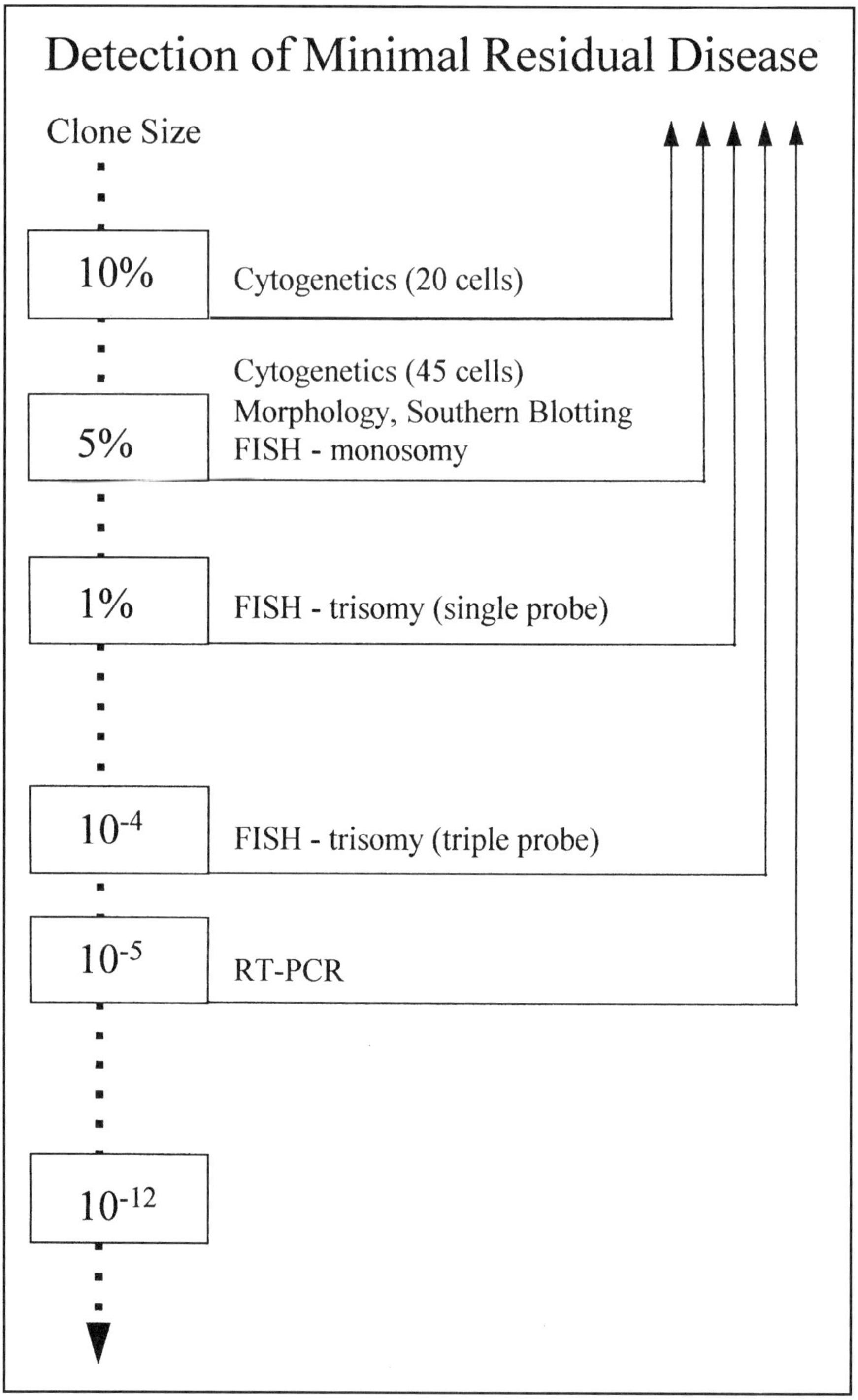

Fig. 9.1. The sensitivity of different techniques for the detection of minimal residual disease. FISH = fluorescence in situ hybridization.

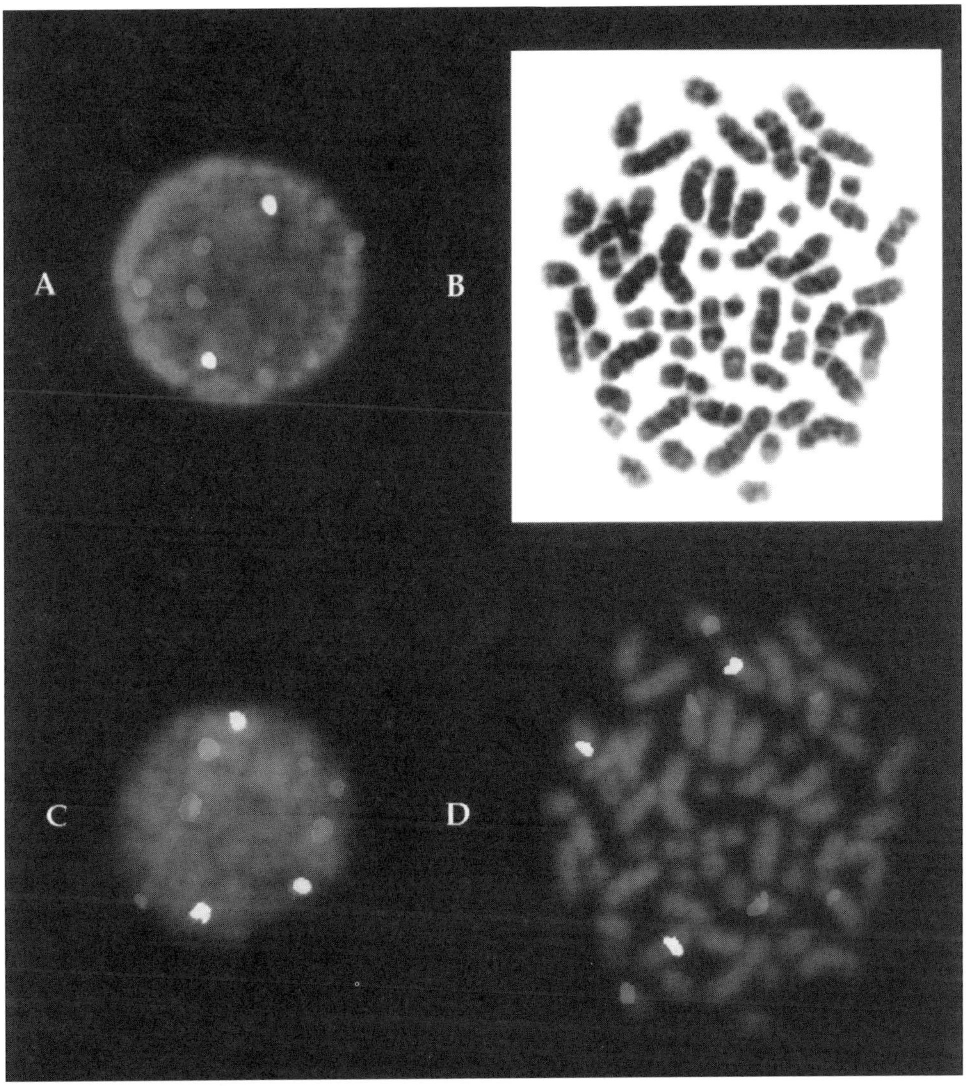

Fig. 9.2 Three color FISH using alpha satellite probes to the centromeric region of chromosomes X (yellow), 6 (red) and 18 (green). A: normal female cell 46, XX, with two copies of each cetromere B: metaphase spread of a hyperdiploid cell with gains of chromosomes X, 6 and 18 showing the G-banding pattern, C: interphase hyperdiploid cell from the same patient showing three copies each of chromosomes X, 6 and 18, D: metaphase spread shown in B showing the probes located at the centromeres. (Kindly donated by Arek Kasprzyk.) See color figure in insert.

Table 9.1 Molecular methods of detecting cytogenetic abnormalities

Abnormality	Genes or DNA sequences	Methods of Detection At diagnosis	in remission
t(1;19)(q23;p13)	*E2A/PBX1*	SB RT-PCR FISH	RT-PCR FISH
t(17;19)(q22;p13)	*E2A/HLF*	SB RT-PCR FISH	RT-PCR FISH
t(4;11)(q21;q23)	*AF4/MLL*	SB RT-PCR FISH	RT-PCR FISH
abn 11q23	*MLL*	SB FISH	FISH
t(9;22)(q34;q11)	*MBCR/ABL*	SB RT-PCR FISH	RT-PCR FISH
	mBCR/ABL	SB RT-PCR FISH	RT-PCR FISH
Monosomy	Chromosome-specific	FISH	FISH
Trisomy	Chromosome-specific	FISH	FISH

SB = Southern blotting
RT-PCR = Reverse transcription polymerase chain reaction
FISH = Fluorescence in situ hybridization

diagnosis when clones are present in >80% of cells. Interphase FISH has been used to corroborate the results of cytogenetic analysis for chromosomal gain in hyperdiploid ALL and to supplement it in patients with failed or normal cytogenetics.[5,6] The detection of trisomies in chromosomally normal samples of interphase cells indicates that hyperdiploidy may be present in the nondividing cell population or in cells which defy analysis.

Moorman et al[7] have shown that the nonrandom nature of chromosome gain in low and high hyperdiploidy enables these clones to be investigated by interphase FISH. No more than four to six chromosome-specific probes were applied in a stepwise fashion. It has been shown that this strategy would enable the detection of 94% of hyperdiploid cases and would predict the ploidy subgroup (low, high or near triploid/tetraploid) in 96% of these patients.

Centromeric Probes for Detection of Chromosomal Loss

The detection of clonal chromosomal loss in interphase is more problematical because control values for monosomy are relatively high. Reports of false positive rates in normal bone marrow have been shown to range from 3.5% for monosomy 8 to 18% for monosomy 9 and even higher rates have been reported.[4] It is therefore crucially important to establish control values for monosomies and trisomies individually for each probe and by each investigator before the technique is applied.

SOUTHERN BLOTTING, RT-PCR AND FISH VS. CYTOGENETICS

The disadvantage of molecular approaches compared with metaphase cytogenetics at diagnosis is that the number of cytogenetic abnormalities which can be detected molecularly is at present limited. Accordingly, the presence or absence of each abnormality must be investigated independently. For translocations it is not yet known how consistent the breakpoints are in all cases. Breakpoints within the *BCR* gene for example occur in three different regions so that detection of the Ph chromosome requires three different sets of primers for *BCR* alone. The development of a multiprimer-PCR is clearly important if there is to be routine scanning for several translocations. This has been described by Izraeli et al[8] and shows the detection of the *E2A/PBX1* transcripts (for t(1;19) and of all three *BCR/ABL* transcripts (for t(9;22)) in a single reaction. For interphase FISH trisomies or monosomies must be individually targeted.

DETECTION OF MINIMAL RESIDUAL DISEASE BY MOLECULAR METHODS

Molecular methods assume particular importance for the detection of residual malignant cells or minimal residual disease (MRD) in remission bone marrows. Indeed in these situations cytogenetics is quite inappropriate because of the low level of sensitivity obtained (see Fig. 9.1). Knowledge of the status of the malignant clone in remission is assuming increasing significance as better treatment protocols are developed and improved survival times are achieved. It is, for example, important to be able to distinguish suppression or partial reduction of the clone (with the associated threat of relapse) from total elimination of the clone which is a more hopeful scenario in terms of cure. The obvious prerequisite for the use of chromosomally-mediated genetic analysis for the detection of MRD is that the clone detected at diagnosis included a cytogenetic abnormality for which probes are available. For the detection of MRD, Southern blotting is not sufficiently sensitive. Remission marrows by definition, contain less than 5% of leukemic blasts and Southern blotting has a similar level of sensitivity. RT-PCR and FISH are both suitable alternatives.

RT-PCR

RT-PCR using a nested approach will discover down to one malignant cell in 100,000 cells. RT-PCR can be used to investigate the presence of the translocations listed above in remission. The prognostic significance of the persistence at low levels of aberrant mRNA from translocations in hematological remission is still under investigation.

FISH

The detection of trisomies in remission marrows has been demonstrated, in several studies, using alpha satellite probes. Hereema et al using a single probe showed persistence of trisomic cells at levels above control values at one month from diagnosis but not later.[5] White et al investigating 13 patients with hyperdiploid ALL detected MRD in 5/13 patients in remission.[9] Kasprzyk et al have shown that single probe FISH for trisomy is relatively insensitive (1%), but for a clone with multiple trisomies the sensitivity of the method can be increased to 10^{-4}. By preparing and scanning cytospins of at least 10,000 cells and applying probes to 3 chromosomes simultaneously, the rare cells with all three trisomies are readily visible (see Fig. 9.2). Heterogeneity in the speed of elimination of the leukemic clone was demonstrated in 16 patients with high hyperdiploidy at diagnosis (13 patients) or in relapse (three patients). Low levels of the clone were detected in between 0.01% and 0.06% of cells in 11/30 samples. In first remission, positive samples were more frequent during the first 7 months of treatment (7/12) than thereafter (2/10). The prognostic significance of this finding awaits further study.[10]

FISH has the advantage that aberrant signals are visualized within individual cells, enabling the observer to concentrate on the malignant cells and to ignore the background of normal cells which show the correct number of the gene(s) or chromosome(s) under review. There has been a report of suspicious cells detected in an early remission marrow of a patient with high hyperdiploid ALL which included trisomy 17. The persistence of the clone was concluded from the detection of trisomy 17 in 106/1,000 cells. When only the suspicious cells were examined, 13/18 (72%) were trisomic further increasing confidence in the observation of delayed remission.[11]

WHOLE CHROMOSOME PAINTING AND FISH

Banding Revisited

Ambiguous karyotypes or metaphases with poor chromosome morphology may benefit from the use of whole chromosome painting probes. The interpretation of chromosomal rearrangements using banding techniques is limited. The banding of normal chromosomes results in alternating light and dark bands of varying size and frequency along the length of the chromosome which is characteristic for each chromosome pair. The number of visible bands decreases with the degree of condensation of the chromosomes and in malignant cells this is frequently considerable. Small segments of dark or light staining regions of DNA on different chromosomes are indistinguishable from each other. In some cases, rearranged chromosomes are undoubtedly misinterpreted as normal. In others, rearrangements of clearly abnormal chromosomes can not be identified. Chromosome painting or the hybridization of whole chromosome probes to a metaphase spread with poor chromosome morphology may be used to confirm or correct a tentative interpretation of a difficult karyotype. The development of DNA probes to whole chromosomes has provided a reliable method of identifying certain derived chromosomes which defy analysis by banding techniques. The use of FISH with whole chromosome probes has the advantage that each chromosome investigated can be labelled with a different color, allowing unequivocal description of the chromosomes contributing to a rearrangement. Although still technically laborious, it is possible to use probes to all 22 autosomes and to the sex chromosomes X and Y. Thus the full karyotype of the most complex clones can be described. The limiting factor in diagnostic patient samples in ALL is the number of metaphases obtained. Whole chromosome FISH has been used in a number of instances to correct or supplement banding techniques and improve the accuracy of the analysis.

An already classic example of the limitations of banding is t(12;21)(p11/13;q22). This translocation cannot be detected on G-banding but was first demonstrated by chromosome painting.[12] Other examples are the two reports of cases with a supposed t(1;19), one of which was shown not to involve fusion of *E2A/PBX1*. On painting with whole chromosome paints, both cases proved to be otherwise. Both were found to have t(1;22). On banding, chromosome 22 had been mistaken for chromosome 19.[13,14]

Whole chromosome paints may be used on interphase cells to show translocations of fairly large chromosome regions. For example, the use of FISH for detecting chromosome 8 translocations in patients with mature B ALL was demonstrated by Nishida et al. A split signal for chromosome 8 in interphase cells was found in 9/15 patients with ALL or non-Hodgkin lymphoma (NHL) but only seven of these showed a split signal in metaphases. In a second study, the same authors showed 6/15 patients to have a split chromosome 8 signal on interphase cells and only two of these to have the same result on metaphases. A third patient had a split 8 signal on metaphase cells which was not seen on interphase.[15]

The centromeric involvement of four patients with hematological malignancies and whole arm translocations of chromosome 7 and 12 has been investigated by Johansson et al. The one patient with ALL, a three year old boy with C-ALL, was shown to have dic(7;12)(q10;q10).[16] Chromosome-specific, alpha satellite probes were used to demonstrate that the centromeres of both chromosomes were present.

YACS AND COSMID PROBES FOR THE DETECTION OF TRANSLOCATIONS AND GENE LOSS OR GAIN

YACs and cosmids containing the affected gene or genes spanning breakpoints are used to discover translocation partners or to confirm deletions. Probes to the immunoglobulin heavy chain gene *(IGH)* have been used to detect translocation partners to 14q32 in patients with B cell malignancies[17] and also to detect the translocations in interphase cells.[18] The use of a YAC containing the *BCR* gene together with a cosmid probe flanking *ABL* to detect the Philadelphia translocation in interphase cells from patients with ALL has been demonstrated.[19] Variant Philadelphia translocations may also be investigated in this way.[20] The first demonstration of chromosomally normal, Philadelphia positive ALL was made using FISH with simultaneous hybridization of probes to *BCR* and *ABL* labelled in different colors. The juxtaposition of the two genes on the apparently normal chromosome 22 was shown (see Fig. 9.2).[21] This raised the possibility that discrepancies between the incidence of Ph+ve patients identified cytogenetically and molecularly may be due, at least in part, to a greater incidence of hidden Ph translocations than previously suspected.

YACs containing sequences to chromosome 6 and 21 have been used to confirm or extend information gathered by cytogenetic analysis on 17 patients with ALL. Of these, 9/17 had a high hyperdiploid clone, five had normal karyotypes and three had abnormal clones without trisomy 6 or 21. In most instances there was concordance between the cytogenetic and FISH results. In two hyperdiploid cases, however, FISH revealed a subclone or a chromosomally normal clone which had not been detected cytogenetically and in another case FISH showed that a marker chromosome included some DNA from chromosome 6.[22]

MORPHOLOGY OR IMMUNOPHENOTYPING WITH FISH

Centromeric probes for trisomies in bone marrow interphase cells can be used, together with May-Grunwald giemsa stain, in order to determine the morphology of the malignant cells and hence the nature of the target cell. In one such study, a 25 year old man with T ALL and t(1;14)(p34;q11) with trisomy 8 was seen to have trisomy 8 in lymphoid blast cells but not in cells belonging to other lineages, thereby confirming the lymphoid commitment of the target cell for chromosomal change.[23] Identification of cell lineages involved in the leukemic clone can also be determined by combining the use of centromeric probes with lineage-specific monoclonal antibodies.

THE FUTURE

The use of molecular techniques to supplement or complement cytogenetic analysis is expanding rapidly. It is likely that probes will soon be available for all nonrandom chromosomal abnormalities and that strategies for their detection by molecular techniques will be perfected.

It is unlikely that chromosome analysis of patients with malignant disease will ever be totally replaced by molecular methods. One of the great conundrums of the cytogenetics of ALL is the extraordinary heterogeneity of chromosomal change found at diagnosis. Most, if not all, of the abnormalities will possibly be found to play a crucial role in leukemogenesis and all chromosomal change presumably makes some contribution to the initiation or progression of malignant change. The genetic mechanisms of many changes at present defy analysis. Nor do we know the prognostic implication of many of these changes or their role in the biology of the disease. At the present time, with improved cytogenetic techniques, the use of image recognition systems and increasing numbers of trained cytogeneticists, chromosome analysis in ALL is a reliable and relatively simple procedure. Within 10 days of diagnosis the clinician can have to hand a cytogenetic profile of the clone. This may decide the most appropriate treatment, indicate prognosis and determine the choice of probes for MRD detection. Further development of the relationship between cytogenetics and molecular biology will continue to play a vital part in the management of patients with acute lymphoblastic leukemia. It may seem possible for treatment to be targeted directly at the aberrant gene products in a way which will destroy the leukemic clone and spare the normal cells. The goal must now be to develop these techniques so that all patients who suffer from acute lymphoblastic leukemia will achieve a complete and lasting recovery.

SUMMARY

In summary, the application of molecular techniques for the detection of cytogenetic abnormalities have extended the role of cytogenetics in a number of ways. It is now possible to detect certain chromosomal abnormalities in DNA or RNA prepared from leukaemic tissue, providing a fall-back position in cases where cytogenetic analysis has not been possible. RT-PCR has a much greater sensitivity than cytogenetics which makes it applicable for studies of minimal residual disease. FISH can be used to identify chromosomal abnormalities in which banding techniques give ambiguous results and to discover the proximity of certain genes to chromosomal breakpoints.

REFERENCES

1. Rowley JD. Human oncogene locations and chromosome aberrations. Nature 1983; 301:290-291.
2. Rabbitts TH. Chromosomal translocations in human cancer. Nature 1994; 372:143-149.
3. Trask BJ. Fluorescence in situ hybridization: applications in cytogenetics and gene mapping. Trends Genet 1991; 7:149-154.
4. Chen Z, Morgan R, Berger CS et al. Application of fluorescence in situ hybridization in hematological disorders. Cancer Genet Cytogenet 1992; 63:62-69.
5. Heerema NA, Argyropoulos G, Weetman R et al. Interphase in situ hybridization reveals minimum residual disease in early remission and return of the diagnostic clone in karyotypically normal relapse of the acute lymphoblastic leukemia. Leukemia 1993; 7:537-543.
6. Berger A, Strehl S, Hekele A et al. Interphase cytogenetic study of childhood acute lymphoblastic leukemia. Medical & Pediatric Oncology 1994; 23:413-421.
7. Moorman AV, Clark R, Farrell DM et al. Probes for hidden hyperdiploidy in acute lymphoblastic leukaemia. Genes Chromosom Cancer 1996; 16:40-45.
8. Izraeli S, Lion T. Multiprimer-PCR for screening of genetic abnormalities in acute lymphoblastic leukaemia. Br J Haematol 1991; 79:645-647.
9. White DM, Crolla JA, Ross FM. Detection of minimal residual disease in childhood acute lymphoblastic leukaemia using fluorescence in-situ hybridization. Br J Haematol 1995; 91:1019-1024.

10. Kasprzyk A, Secker-Walker LM. Minimal residual disease detection by multiprobe FISH in acute lymphoblastic leukaemia with hyperdiploidy. Br J Haematol 1996; 93:58.
11. Anastasi J, Vardiman JW, Rudinsky R et al. Direct correlation of cytogenetic findings with cell morphology using in situ hybridization: an analysis of suspicious cells in bone marrow specimens of two patients completing therapy for acute lymphoblastic leukemia. Blood 1991; 11:2456-2462.
12. Romana SP, Le Coniat M, Berger R. t(12;21): a new recurrent translocation in acute lymphoblastic leukemia. Genes Chromosom Cancer 1994; 9:186-191.
13. Rieder H, Kolbus U, Koop U et al. Translocation t(1;22) mimicking t(1;19) in a child with acute lymphoblastic leukemia as revealed by chromosome painting. Leukemia 1993; 7:1663-1666.
14. Filatov LV, Behm FG, Pui CH et al. Childhood acute lymphoblastic leukemia with equivocal chromosome markers of the t(1;19) translocation. Genes Chromosom Cancer 1995; 13:99-103.
15. Nishida K, Ritterbach J, Repp R et al. Characterization of chromosome 8 abnormalities by fluorescence in situ hybridization in childhood B-acute lymphoblastic leukemia/non-Hodgkin lymphoma. Cancer Genet Cytogenet 1995; 79:8-14.
16. Johansson B, Arheden K, Hoglund M et al. Fluorescence in situ hybridization analysis of whole-arm 7;12 translocations in hematologic malignancies. Genes Chromosom Cancer 1995; 14:56-62.
17. Taniwaki M, Nishida K, Ueda Y et al. Interphase and metaphase detection of the breakpoint of 14q32 translocations in B-cell malignancies by double-color fluorescence in situ hybridization. Blood 1995; 85:3223-3228.
18. Taniwaki M, Matsuda F, Jauch A et al. Detection of 14q32 translocations in B-cell malignancies by in situ hybridization with yeast artificial chromosome clones containing the human IgH gene locus. Blood 1994; 83:2962-2969.
19. Bentz M, Cabot G, Moos M et al. Detection of chimeric BCR-ABL genes on bone marrow samples and blood smears in chronic myeloid and acute lymphoblastic leukemia by in situ hybridization. Blood 1994; 83:1922-1928.
20. Nakai H, Taniwaki M, Tanaka S et al. Characterization of a 14q+ marker chromosome in philadelphia chromosome positive acute lymphoblastic leukemia by DNA analysis and fluorescence in situ hybridization. [Review]. Cancer Genet Cytogenet 1995; 81:83-91.
21. Van Rhee F, Kasprzyk A, Jamil A et al. Detection of the BCR-ABL gene by reverse transcription/polymerase chain reaction and fluorescence in situ hybridization in a patient with Philadelphia

chromosome negative acute lymphoblastic leukaemia. Br J Haematol 1995; 90:225-228.

22. Romana SP, Cherif D, Le Coniat M et al. In situ hybridization to interphase nuclei in acute leukemia. Genes Chromosom Cancer 1993; 8:98-103.
23. Martin-Henao GA, Sureda A. Translocation (1;14)(p34;q11) and trisomy 8 in a T-cell acute lymphoblastic leukemia patient. Cancer Genet Cytogenet 1995; 79:177-181.

APPENDIX 1

A Glossary of Cytogenetic Terms

Aneuploidy	Deviation of the chromosome number from that characteristic for the species. In the human, a chromosome complement that is not an exact multiple of the haploid (n) number (23 chromosomes).
Autosome	Any chromosome other than the sex chromosomes.
Banding	The resolution of the various intrachromosomal bands and/or regions of varying intensity by the procedure of differential staining known as G-banding.
Breakpoint	Location of a break in a chromatid or chromosome denoted by the band involved.
Centromere	Area of chromosomal constriction, especially visible in metaphase, where the two chromatids are held together and are attached to the spindle. Another term for the centromere is *primary constriction*.
Chromatid	Two sister chromatids, held together at the centrometric region, make up a chromosome and, hence, each sister chromatid represents the replicated form of the genetic material of the other.

	One chromatid of each chromosome migrates to each new cell when cellular division is completed.
Chromatin	Substance of which chromosomes are composed.
Chromosome	Arrangement of the nuclear genetic material (chromatin, DNA) into small and discrete bodies, typically containing a centromere and two chromosome arms, as seen at metaphase during cell division. Chromosomes contain the genes, or hereditary factors, and are constant in number in each species. The normal number in humans is 46, with 22 pairs of autosomes and 2 sex chromosomes (XX or XY).
Clonal evolution	Continuous stepwise evolution of new chromosomal patterns by successive acquisitions, losses, and/or duplications of whole chromosomes or their parts and the development of rearrangements.
Clone	A clone is the term used when two or more cells have the same structural chromosome change or extra chromosome, or when at least three cells with the same missing chromosome are found.
Cytogenetics	Study of chromosomes.
Deletion	Loss of a chromosome segment.
Derivative chromosome	An abnormal chromosome resulting from structural rearrangement involving two or more chromosomes.
Dicentric	A chromosome having two centromeres.

Diploidy ($2n$) State of having two full sets of homologous chromosomes. Each species has a characteristic number ($2n$), with the haploid number (n) being present in the gamete (sperm or ovum). In humans, $2n$ = 46.

Distal Closer to the telomere than to the centromere.

Double minute chromosome Small acentric and paired fragments of somewhat variable size.

Endo-reduplication Endoreduplication of chromosomes occurs during interphase and results in the formation of metaphase chromosomes with four sister chromatids (diplochromosomes) and exact karyotype doubling. Selective endoreduplication of a single chromosome, for example, Ph, or parts of chromosomes has been postulated.

Fragment Broken part of a chromosome: in an acentric fragment, no centromere is present, whereas a centric fragment contains a centromere.

Genome Total genetic constitution of a cell or organism.

Haploidy State, usually of germ cells, of having n chromosomes. In the human, n = 22 + X in the mature ovum and 22+ Y or X in the mature sperm.

Hetero-chromatin Parts of a chromosome may appear condensed. Using differential staining, condensed regions stain dark and are called heterochromatin. *Constitutive* heterochromatin is permanent and is present in all human chromosomes in variable quantities. In

	humans, it is localized in the centromeric region, in the secondary constrictions of chromosomes 1, 9, and 16 on their long arm and the distal part of the long arm of the Y.
Homogeneously staining region	Abnormal region in a chromosome lacking the longitudinal differentiation ordinarily revealed by banding.
Hyperdiploid	Containing more than the diploid number of chromosomes; in the human, more than 46.
Hypodiploid	Containing fewer than the diploid number of chromosomes; in the human, less than 46.
Idiogram	Diagrammatic representation of a karyotype.
Insertion	Transfer of a chromosome segment into another chromosome; involves two breakages in one chromosome, and insertion of the broken segment into a breakage point in another.
Interphase	Portion of the cell cycle when the cell is not in mitosis. Interphase includes the G_1, S, and G_2 phases. During interphase, the chromosomes are active in RNA synthesis. At a certain, rather fixed time before the cell next divides, the chromosomes replicate themselves and thus double the amount of chromosomal material in the nucleus. The time at which they are doing this is known as the S (synthesis) period.
Interstitial deletion	Loss of chromosomal material not involving the terminal ends of the chromosome.
Inversion	Result of two breaks in the same chromosomes with rotation of the segment between them 180°

	and fusion of the broken ends. In *paracentric inversion*, the breaks occur on the same arm on one side of the centromere; the intraarm exchange may lead to no apparent altered morphology. A *pericentric inversion* is one that occurs as a result of breakage followed by relocation of a segment within a chromosome about the centromere.
Isochromatid break or deletion	Discontinuity in both chromatids of a chromosome at identical sites.
Isochromosome	Symmetric chromosome composed of duplicated long or short arms formed after misdivision of the centromere in a transverse plane.
Karyogram	Pictorial display of chromosomes from a cell, paired and arranged in descending order of size.
Karyotype	Written description of the chromosomal makeup of the cell.
Kinetochore	Discrete structure positioned along the surface of the centromere that functions as the point of attachment of the mitotic spindle for the chromosome. A chromosome may have more than one centromere, but usually has only one kinetochore.
Marker chromosome	Abnormal chromosome of unknown origin(s).
Metacentric	Describes a chromosome with the centromere located midway between the two extremities.
Metaphase	When chromosomes are maximally condensed, most easily seen, and least genetically active, and

	they become attached to the spindle at the centromere by means of the kinetochore.
Mitosis	Process of cellular division resulting, under normal conditions, in two daughter cells with the diploid number of chromosomes.
Modal chromosome number	Predominant number in a population of cells with a range of chromosome numbers.
Monosomic	Having only one chromosome of a homologous pair, instead of the normal two.
Monosomy	Absence of one member of a homologous pair of chromosomes.
n	Haploid number of chromosomes; in the human, 23.
$2n$	Diploid number of chromosomes; in the human, 46.
Nondisjunction	Failure of two daughter chromosomes to separate at anaphase. Thus, both migrate to the same pole along the spindle, and one daughter nucleus contains more, and the other fewer, chromosomes than normal.
Oncogene	Gene involved in the process of malignant transformation.
Ploidy	State of having a specific multiple of a set of chromosomes or deviations from it, that is, diploid ($2n$), triploid ($3n$), tetraploid ($4n$) and so forth.

Polymorphism — Presence in a population of coexistent alternative phenotypes. Applied in cytogenetics to differences in chromosome morphology, particularly banding patterns.

Polyploidy — State of having a multiple of *n* chromosomes but more than 2*n*; in the case of a tumor, a multiple of the modal number of chromosomes.

Prophase — First stage of cell division; initiated when the chromosomes begin to become visible in the nucleus as fine threads. The chromosomes progressively shorten, thicken, and become more obvious as prophase proceeds.

Proto-oncogene — Inactive formof an oncogene.

Proximal — Closer to the centromere than to the telemere.

Pseudodiploidy — In humans, the state of having 46 chromosomes but with an abnormal karyotype.

Ring chromosome — Chromosome in which two breaks occur on opposite sides of the centromere; the ends of the centric fragment fuse to form a centric ring chromosome.

Sex chromosome — In humans, the X and Y chromosomes.

S period — Interval, between the G_1 and G_2 periods, during which the chromosomes replicate their DNA.

Stemline — Number of chromosomes characterizing the modal cells in cancer or leukemia.

Telomere — Terminal region of a chromosome.

Telomeric fusion	Fusion of the ends (telomeres) of two chromosomes to form a compound, possibly dicentric chromosome.
Telophase	Last stage of cell division, in which the cytoplasm undergoes cleavage to form two completely separate daughter cells.
Terminal deletion	Deletion that occurs near a terminal region of the chromosome.
Tetraploidy	State of having twice ($4n$) the number of chromosomes in diploid cells; in the human, 92. See Ploidy.
Translocation	Breakage followed by transfer of chromosome material between chromosomes. Translocation is *reciprocal* when it is balanced; if material is lost, it is unbalanced. Central fusion (*Robertsonian*) types of translocation form abnormal chromosomes by fusion of two acrocentric chromosomes in the centromeric region.
Triploidy ($3n$)	State of having $3n$ chromosomes; in the human, 69.
Trisomy	Presence of an extra (third) autosome in addition to a normal homologous pair. Also, presence of an extra sex chromosome.

Adapted from Sandberg A. The Chromosomes in Human Cancer and Leukemia. 2nd Ed. New York: Elsevier, 1990.

APPENDIX 2

Leukemia Research Fund/ United Kingdom Cancer Cytogenetics Group Karyotype Database in Acute Lymphoblastic Leukemia

Director: Lorna M. Secker-Walker

The LRF/UKCCG Karyotype database in ALL is a repository of the cytogenetic findings at diagnosis of patients with ALL entered to the Medical Research Council's treatment trials for infants (aged <1 year), children (aged 1-14 years) and adults (15-55 years). It was started in March 1992 and by June 1996 it comprised successful karyotypic analyses from 26 infants, 1186 children and 156 adults. Details such as the following which appear in some of the tables are regularly updated and circulated to contributing Cytogeneticists and Haematologists in the UK.

Table A1. Classification of 1186 children and 156 adults with ALL by ploidy

Age group in years	Total cases	normal 46	Ho <46	ps 46R	hel 47-50	heh 51-65	trip 66-80	tetra >80
1-14	1186	211	81	361	142	366	10	15
15-55	156	30	12	73	13	21	3	4

Table A2. Classification of children and adults by structural change

Age group in years	total no.	t(1;19) ALL total no.	C-ALL 1;19/ total*	Pre-B 1;19/ total*	t(4;11) no.	t(9;22) no.	dup1q no.
1-14	1186	32	12/592	14/113	10	20	46
15-55	156	3	1/57	1/14	12	21	1

* with known immunophenotype

Table A3. Clinical features of patients with t(4;11)

Age group in years	total	M:F	t(4;11) Age in years median (range)	$WBC X 10^9/L$ median (range)
1-14	10	6:4	5.5 (1-14)	43 (3-535)
15-55	12	2:10	40 (15-55)	247 (7-780)

Table A4. Incidence of children with abn6q and abn9p with B-lineage and T ALL.

Age group in years	total cases	ALL abn6q no.	abn6q B-lin abn6q/ total*	T ALL abn6q/ total*	ALL abn9p	abn9p B-lin abn9p/ total*	T ALL abn9p/ total*
1-14	1186	82	49/753	9/110	67	42/753	10/110

* with known immunophenotype

Table A5. Cytogenetic classification of 26 infants by age group

Age group	total	normal	t(4;11)	t(9;11)	t(11;19)	other abn11q	other
<6 months	11	1	6	1	0	2	1
>6 months	15	2	5	1	0	1	6
Sex M:F	11:15	1:2	4:7	0:2	0	0:3	6:1

INDEX

R

S

T

U

Y

THE LIBRARY
IC SCHOOL OF MEDICINE
AT ST. MARY'S

THE LIBRARY
IC SCHOOL OF MEDICINE
AT ST. MARY'S